KU-548-157

Nankai Tracts in Mathematics – Vol.1

SCISSORS CONGRUENCES,
GROUP HOMOLOGY

AND

CHARACTERISTIC CLASSES

Johan L. Dupont

University of Aarhus, Denmark

World Scientific
Singapore • New Jersey • London • Hong Kong

Published by

World Scientific Publishing Co. Pte. Ltd.

P O Box 128, Farrer Road, Singapore 912805

USA office: Suite 1B, 1060 Main Street, River Edge, NJ 07661

UK office: 57 Shelton Street, Covent Garden, London WC2H 9HE

British Library Cataloguing-in-Publication Data

A catalogue record for this book is available from the British Library.

SCISSORS CONGRUENCES, GROUP HOMOLOGY AND CHARACTERISTIC CLASSES

Copyright © 2001 by World Scientific Publishing Co. Pte. Ltd.

All rights reserved. This book, or parts thereof, may not be reproduced in any form or by any means, electronic or mechanical, including photocopying, recording or any information storage and retrieval system now known or to be invented, without written permission from the Publisher.

For photocopying of material in this volume, please pay a copying fee through the Copyright Clearance Center, Inc., 222 Rosewood Drive, Danvers, MA 01923, USA. In this case permission to photocopy is not required from the publisher.

ISBN 981-02-4507-6
ISBN 981-02-4508-4 (pbk)

Printed in Singapore by FuIsland Offset Printing

Preface

For almost 20 years Chih-Han Sah and I worked together on various aspects of Hilbert's Third Problem and in particular its relation to homological algebra and algebraic K-theory. We often talked about writing a book on this subject and the development in this area since Han's book from 1979, but we always got sidetracked by some interesting problem and the plans never materialized. After Han's untimely death in 1997 his large collection of mathematical books were donated by Analee Sah to the Nankai Institute of Mathematics in Tenjin and professor S.-S. Chern kindly asked me to visit this institute and to give a series of lectures on the above topic. Professor Chern also suggested that I write up these lectures which I gave during my visit to Nankai University in the fall of 1998. These notes are far from the grand project which Han and I had talked about. But I hope they will give an impression of the great variety of mathematical ideas which are related to the seemingly elementary subject of scissors congruence.

I am very grateful to professor Chern and director X. Zhou of the Nankai Institute for a wonderful stay in China. Also I would like to thank the audience to my lectures, especially J. Pan, W. Zhang and F. Fang for their interest and kindness. Furthermore I want to thank M. Bökstedt, J.-L. Cathelineau, J.-G. Grebet and F. Patras for valuable comments and corrections. Special thanks are due to Karina Thorup Mikkelsen for typing the manuscript of these notes and to S. Have Hansen for helping with the drawings and other Latex problems. Finally I would like to acknowledge the support of the Danish Natural Science Research Council and Aarhus University.

Aarhus, August, 2000,   Johan Dupont

Contents

Preface   v

Chapter 1.   Introduction and History   1

Chapter 2.   Scissors congruence group and homology   9

Chapter 3.   Homology of flag complexes   17

Chapter 4.   Translational scissors congruences   27

Chapter 5.   Euclidean scissors congruences   37

Chapter 6.   Sydler's theorem and non-commutative differential forms   45

Chapter 7.   Spherical scissors congruences   53

Chapter 8.   Hyperbolic scissors congruence   63

Chapter 9.   Homology of Lie groups made discrete   77

Chapter 10.   Invariants   91

Chapter 11.   Simplices in spherical and hyperbolic 3-space   107

Chapter 12.   Rigidity of Cheeger-Chern-Simons invariants   119

Chapter 13.   Projective configurations and homology of the
              projective linear group   125

Chapter 14.   Homology of indecomposable configurations   135

Chapter 15.   The case of $\mathrm{PGl}(3, F)$ 145

Appendix A.   Spectral sequences and bicomplexes   151

Bibliography   159

Index   167

CHAPTER 1

Introduction and History

It is elementary and well-known that two polygons P and P' in the Euclidean plane have the same area if and only if they are *scissors congruent (s.c.)*, i.e. if they can be subdivided into *finitely* many pieces such that each piece in P is congruent to exactly one piece in P'. In this form the problem was explicitly stated and solved by W. Wallace (cf. [**Wallace, 1807**], [**Jackson, 1912**]) but I believe it was known to the ancient Greeks or Chinese. In fact it originates from an attempt to give an elementary definition of "area": Given a line segment of unit length "1" one wants the following.

THEOREM 1.1. *Any plane polygon P is s.c. to a rectangle with one side of length 1. Hence the length of the other side measures the area.*

Proof: By triangulation of P it is enough to take P a triangle. By the following figure it is s.c. to a parallelogram:

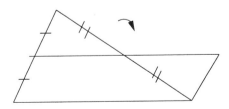

Another figure shows a s.c. to a rectangle

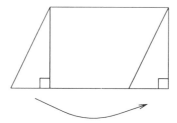

However using this step backwards we can adjust the distance between the "other" parallel sides until it is of unit length.

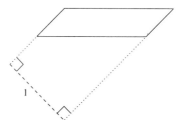

Finally use the previous step to obtain the desired rectangle.   □

As for the corresponding problem in Euclidean 3-space, C. F. Gauss mentions in a letter [**Gauss, 1844**] that the proof that two pyramids with the same base and height have the same volume, uses "exhaustion", i.e. subdivisions of the polyhedra into *infinitely* many pieces, and he asked for a proof using only *finitely* many pieces.

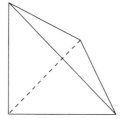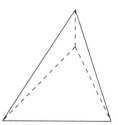

Two pyramids with the same base and height.

However Hilbert did not believe that this was possible and suggested as the 3rd problem on his famous list presented at the International Congress in Paris 1900 the following: Find two polyhedra with the same volume and show that they are *not* s.c. This was in fact done already in the same year by M. Dehn (see [**Dehn, 1901**]) who showed that the regular cube and the regular thetrahedron of the same volume are *not* s.c. A somewhat incomplete proof along the same lines was already published by R. Bricard [**Bricard, 1896**]. In modern formulation Dehn's proof is most elegantly given in terms of the *Dehn invariant* D which takes values in the tensor product of abelian groups $\mathbb{R} \underset{\mathbb{Z}}{\otimes} (\mathbb{R}/\mathbb{Z})$. This is defined for a polyhedron P by the formula

$$(1.2) \qquad D(P) = \sum_A \ell(A) \otimes (\theta(A)/\pi)$$

where A runs through the collection of all edges of P and where $\ell(A)$ is the length of A and $\theta(A)$ is the dihedral angle of P at A.

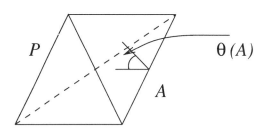

Dehn did not know tensor products (introduced by H. Whitney in [**Whitney, 1938**]); he argued directly on the relations between the occurring side lengths and dihedral angles. In terms of the Dehn invariant we can rephrase Dehn's theorem as follows:

THEOREM 1.3.   a) *If P and P' are s.c. polyhedra in Euclidean 3-space then $D(P) = D(P')$.*
   b) *If C is a cube and T is a regular tetrahedron then $D(C) = 0$ whereas $D(T) \neq 0$.*

Proof: a) can be proved directly (for a proof see e.g. [**Boltianskii, 1978**] or [**Sah, 1979**]). But it also follows from the results of chapter 5 below.

b) Since in C all dihedral angles are $\pi/2$ and thus a rational multiple of π it follows that $D(C) = 0$. As for the regular tetradron T elementary trigonometry yields that all dihedral angles are $\varphi = \mathrm{Arc\,cos}\,1/3$, so that

$$D(T) = 6\ell \otimes \varphi/\pi$$

where ℓ is the common length of the edges. Considering $\mathbb{R}$ as a rational vector space we just have to show that φ/π is irrational. For this we use the formula

$$\cos(k+1)\varphi + \cos(k-1)\varphi = 2\cos k\varphi \cos\varphi$$

to show by induction on $k \geq 1$ that $\cos k\varphi = a_k/3^k$ where a_k is an integer *not* divisible by 3. $\square$

It is now an obvious question if the s.c. class of a polyhedron is determined by its volume and Dehn invariant. This was answered affirmatively in 1965 by J. P. Sydler who proved the following (see [**Sydler, 1965**] or [**Jessen, 1968**]) :

THEOREM 1.4. *If two Euclidean polyhedra P and P' satisfy* $\mathrm{Vol}(P) = \mathrm{Vol}(P')$ *and* $D(P) = D(P')$ *then P and P' are s.c.*

We shall sketch a proof of this using homological algebra in chapter 6. In fact, it is the main theme of these lectures to show the relationship between Hilbert's 3rd Problem and questions in modern homological algebra and topology. First let us formulate the problem more generally:

For $n \geq 1$ let $X = X^n = E^n, S^n, \mathcal{H}^n$ denote respectively Euclidean, spherical or hyperbolic n-space. Also let $I(X)$ denote the group of isometries of X. A *geometric n-simplex* $\Delta \subseteq X$ is the geodesic convex hull $\Delta = |(a_0, \cdots, a_n)|$ of $n+1$ points $a_0, \cdots, a_n \in X$ in general position (i.e. *not* lying on a hyperplane). In the spherical case we assume that $a_0, \cdots, a_n$ all lie in an open hemi-sphere. The points $a_0, \cdots, a_n$ are the *vertices* of Δ. A polytope $P \subseteq X$ is any finite union $P = \bigcup_{j=1}^k \Delta_j$ of simplices such that $\Delta_i \cap \Delta_j$ is a common face of lower dimension if $i \neq j$. If P, P_1, P_2 are polytopes such that $P = P_1 \cup P_2$ and $P_1 \cap P_2$ has no interior points then we will say that P *decomposes* into P_1 and P_2 and we shall write $P = P_1 \amalg P_2$.

DEFINITION 1.5. Let $G \subseteq I(X)$ be a subgroup. Two polytopes P, P' are called *G-scissors congruent* denoted $P \underset{G}{\sim} P'$ if $P = \amalg_{i=1}^{k} P_i$, $P' = \amalg_{i=1}^{k} P_i'$ and $P_i = g_i P_i$ for some $g_1, \cdots, g_k \in G$.

Generalized Hilbert's 3rd Problem. Find computable invariants which determine when two polytopes are G-s.c.

We shall mainly take $G = I(X)$ and just write $\sim$ for $\underset{I(X)}{\sim}$. In view of Sydler's theorem it is natural to ask the following question raised by B. Jessen [**Jessen, 1973**];
[**Jessen, 1978**]:

Non-Euclidean Hilbert's 3rd Problem. Do volume and Dehn invariant determine the s.c. classes of polyhedra in spherical or hyperbolic 3-space?

Let us reformulate the general problem in *algebraic terms*

DEFINITION 1.6. The *scissors congruence group* $\mathcal{P}(X, G)$ (for $G = I(X)$ write $\mathcal{P}(X)$) is the free abelian group on symbols $[P]$ for all polytopes P in X, modulo the relations:

i) $[P] - [P'] - [P'']$ for $P = P' \amalg P''$,
ii) $[gP] - [P]$ for $g \in G$.

Note that $[P] = [P']$ in $\mathcal{P}(X, G)$ if and only if P and P' are stably G-s.c., that is if there exist polytopes Q and Q' such that $P \amalg Q \underset{G}{\sim} P' \amalg Q'$ with $Q' = gQ$ for some $g \in G$.

However if G acts transitively on X it is a theorem of V. B. Zylev [**Zylev, 1965**] (see e.g. [**Sah, 1979**] for a proof) that stably G-s.c. implies G-s.c. Hence the Generalized Hilbert's 3rd Problem can be reformulated as the problem of determining the s.c. groups $\mathcal{P}(X, G)$. We shall see that this is closely related to the problem of determining the *homology* of the group G:

Recall that for any group G and M a (left) G-module the *group homology* $H_n(G, M)$ is the nth homology group of the chain complex $C_*(G, M)$ which in degree n has generators of the form

$$[g_1|\cdots|g_n] \otimes x, \quad g_1, \cdots, g_n \in C, \quad x \in M$$

and where the boundary map $\partial : C_n(G, M) \to C_{n-1}(G, M)$ is given by

$$\partial([g_1|\cdots g_n] \otimes x) = [g_2|\cdots|g_n] \otimes g_1^{-1}x + \sum_{i=1}^{n-1}(-1)^i[g_1|\cdots g_i g_{i+1}|\cdots g_n] \otimes x$$
$$+ (-1)^n[g_1|\cdots|g_{n-1}] \otimes x.$$

That is

$$H_n(G, M) = \frac{\ker[\partial : C_n(G, M) \to C_{n-1}(G, M)]}{\mathrm{im}[\partial : C_{n+1}(G, M) \to C_n(G, M)]}.$$

As an example of the relation between s.c. and homological algebra we shall prove the following result from [**Dupont, 1982**] ($H_*(G) = H_*(G, \mathbb{Z})$):

THEOREM 1.7. *There are exact sequences of abelian groups:*

a)

$$0 \to H_2(\mathrm{SO}(3), \mathbb{R}^3) \xrightarrow{\sigma} \mathcal{P}(E^3)/\mathcal{Z}(E^3) \xrightarrow{D} \mathbb{R} \otimes \mathbb{R}/\mathbb{Z} \to$$
$$\to H_1(\mathrm{SO}(3), \mathbb{R}^3) \to 0$$

b)

$$0 \to H_3(\mathrm{Sl}(2, \mathbb{C}))^- \xrightarrow{\sigma} \mathcal{P}(\mathcal{H}^3) \xrightarrow{D} \mathbb{R} \otimes \mathbb{R}/\mathbb{Z} \to H_2(\mathrm{Sl}(2, \mathbb{C}))^- \to 0$$

c)

$$0 \to H_3(\mathrm{SU}(2)) \xrightarrow{\sigma} \mathcal{P}(S^3)/\mathcal{Z} \xrightarrow{D} \mathbb{R} \otimes \mathbb{R}/\mathbb{Z} \to H_2(\mathrm{SU}(2)) \to 0$$

Notation: In a) $\mathcal{Z}(E^3)$ is the subgroup generated by all prisms (i.e. a product of a line segment and a plane polygon). In b) $^-$ indicates the

(-1)-eigenspace for the involution induced by complex conjugation. In c) $\mathbb{Z} \subseteq \mathcal{P}(S^3)$ is generated by $[S^3]$.

We remark that by theorem 1.1 the volume in the Euclidean case gives an isomorphism $\mathcal{Z}(E^3) \cong \mathbb{R}$. Hence by theorem 1.7 Sydler's theorem is equivalent to $H_2(\mathrm{SO}(3), \mathbb{R}^3) = 0$. We shall return to this in chapter 6. In the non-Euclidean cases b) and c) the volume gives rise to cohomology classes for the discrete groups underlying $\mathrm{Sl}(2, \mathbb{C})$ respectively $\mathrm{SU}(2)$. These are in fact equivalent to a *secondary characteristic class* defined by Cheeger-Chern-Simons (see [**Chern-Simons, 1974**] and [**Cheeger-Simons, 1985**]) for bundles with flat structure groups. We shall investigate this in chapter 10 and 12. Furthermore in chapter 9 we shall see that the homology groups in b) and c) are computable in terms of *algebraic K-theory* of the field $\mathbb{C}$ of complex numbers, and we thereby reduce the non-Euclidean Hilbert's 3rd Problem to a well-known difficult Rigidity Problem.

The last 3 chapters (13-15) continue the study in chapter 9 of the homology of the general linear group in terms of groups similar to scissors congruence groups but based on *projective configurations*.

For further information on the history and background of Hilbert's 3rd Problem we refer to the survey [**Cartier, 1985**] and the books [**Boltianskii, 1978**], [**Hadwiger, 1957**], [**Sah, 1979**].

CHAPTER 2

Scissors congruence group and homology

In this chapter we shall investigate the s.c. group further, and in particular we shall give it a homological description in general. As before let $X = X^n$ be either E^n, S^n or $\mathcal{H}^n$, and let $I(X)$ denote the group of isometries of X. Thus E^n is the affine real n-dimensional space over the vector space $\mathbb{R}^n$ and $I(E^n) = E(n)$ is the Euclidean group $E(n) = T(n) \rtimes O(n)$ where $O(n)$ is the orthogonal group and $T(n) \cong (\mathbb{R}^n, +)$ is the additive group of translations

$$T(n) = \{t_v \mid v \in \mathbb{R}^n \quad ; \quad t_v(x) = x + v, \, x \in \mathbb{R}\}.$$

As usual

$$S^n = \{x \in \mathbb{R}^{n+1} \mid x_0^2 + x_1^2 + \cdots + x_n^2 = 1\}$$

and $I(S^n) = O(n+1)$. Similarly

$$\mathcal{H}^n = \{x \in \mathbb{R}^{n+1} \mid -x_0^2 + x_1^2 \cdots + x_n^2 = -1, \, x_0 > 0\}$$

and $I(\mathcal{H}^n) = O^+(1, n)$, the subgroup of $O(1, n)$ keeping $\mathcal{H}^n$ invariant. Occasionally we shall need also the upper halfspace model or the disk model for the hyperbolic space.

As before let $G \subseteq I(X)$ be a subgroup and $\mathcal{P}(X, G)$ the G-s.c. group for X. The following is obvious from the definition:

PROPOSITION 2.1. a) *Suppose $H \trianglelefteq G$ is an invariant subgroup. Then the canonical map $\mathcal{P}(X, H) \to \mathcal{P}(X, G)$ induces an isomorphism*

$$\mathcal{P}(X, G) \cong H_0(G/H, \mathcal{P}(X, H)).$$

b) *In particular*

$$\mathcal{P}(X, G) \cong H_0(G, \tilde{\mathcal{P}}(X))$$

where $\tilde{\mathcal{P}}(X) = \mathcal{P}(X, \{1\})$.

As an example let us show the following theorem often attributed to C. L. Gerling (see however [**Sah, 1979**, Remark 4.2]): Let $I^+(X) \subseteq I(X)$ be the subgroup of orientation preserving isometries.

THEOREM 2.2. *The natural map*

$$\mathcal{P}(X, I^+(X)) \to \mathcal{P}(X, I(X)) = \mathcal{P}(X)$$

is an isomorphism.

Proof: Since $I^+(X)$ is a normal subgroup of index 2 we get

$$\mathcal{P}(X) = H_0(\mathbb{Z}/2, \mathcal{P}(X, I^+(X)))$$

where the action by the generator g of $\mathbb{Z}/2$ is induced by any reflection in a hyperplane. Since $\mathcal{P}(X, I^+(X))$ is generated by n-simplices it suffices to show for an n-simplex Δ that

(2.3) $$g[\Delta] = [\Delta].$$

Let $\Delta = |(a_0, \ldots, a_n)|$ and let $\{U_{ij}\}_{i<j}$ be the collection of hyperplanes bisecting the dihedral angles at the codimensional 2 faces $|(a_0, \ldots, \hat{a}_i, \ldots, \hat{a}_j, \ldots, a_n)|$. By elementary geometry these hyperplanes intersect in a single point c, the "inscribed sphere center". (Notice that this exists also in spherical and hyperbolic geometry.) Also let c_i be the orthogonal projection of c onto the ith $(n-1)$-face of Δ. The reflection r_{ij} in U_{ij} clearly fixes c and interchanges c_i and c_j so that the two simplices

$$\Delta_{ij} = |(c, c_i, a_0, \ldots, \hat{a}_i, \ldots, \hat{a}_j, \ldots, a_n)|$$

and

$$\Delta_{ji} = |(c, c_j, a_0, \ldots, \hat{a}_i, \ldots, \hat{a}_j, \ldots, a_n)|$$

are congruent (see fig.)

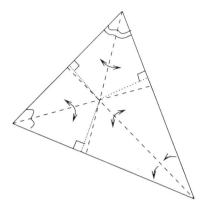

Since

(2.4)
$$\Delta = \coprod_{i<j}(\Delta_{ij} \amalg \Delta_{ji})$$

and

$$g[\Delta_{ij}] = [\Delta_{ji}], \quad i \neq j,$$

we get in $\mathcal{P}(X, I^+(X))$

$$g[\Delta] = \sum_{i<j}(g[\Delta_{ij}] + g[\Delta_{ji}]) = \sum_{i<j}([\Delta_{ji}] + [\Delta_{ij}]) = \Delta$$

so that $\mathbb{Z}/2$ acts trivially on $\mathcal{P}(X, I^+(X))$. $\square$

COROLLARY 2.5. $\mathcal{P}(X)$ is 2-divisible.

Proof: By (2.4) we have in $\mathcal{P}(X)$

$$[\Delta] = \sum_{i<j}[\Delta_{ij}] + [\Delta_{ji}] = 2\sum_{i<j}[\Delta_{ij}]$$

so that $[\Delta]$ is divisible by 2 in $\mathcal{P}(X)$. $\square$

Remark: As we shall see later, $\mathcal{P}(E^n), \mathcal{P}(S^2)$ and $\mathcal{P}(\mathcal{H}^2)$ are vector spaces over $\mathbb{R}$ whereas $\mathcal{P}(S^3)$ and $\mathcal{P}(\mathcal{H}^3)$ are rational vector spaces. It is not known for $\mathcal{P}(S^n)$ and $\mathcal{P}(\mathcal{H}^n), n > 3$, if these groups are p-divisible for $p \neq 2$.

Proposition 2.1 b) is just a reformulation in homological terms of the relation ii) in the definition of $\mathcal{P}(X, G)$. We shall now see that also the relation i) is a homological one. In fact we shall interpret $\tilde{\mathcal{P}}(X)$ as a homology group. Let $C_*(X)$ be the chain complex where a generator or *simplex* in degree k is a $(k+1)$-tuple $\sigma = (a_0, \ldots, a_k)$ of points in X. In the case of $X = S^n$ we assume furthermore that all $(a_0, \ldots, a_k)$ are contained in an open hemi-sphere. For $k = n$ and $a_0, \ldots, a_n$ in general position σ defines a geometric n-simplex $|\sigma|$ as the geodesic convex hull of the vertices, and σ is called *proper* in this case. The boundary map in $C_*(X)$ is just given by

$$(2.6) \qquad \partial(a_0, \ldots, a_k) = \sum_{i=0}^{k} (-1)^i (a_0, \ldots, \hat{a}_i, \ldots, a_k)$$

Now the idea is that for $k = n + 1$ the identity

$$(2.7) \qquad \sum_{i=0}^{n+1} (-1)^i (a_0, \ldots, \hat{a}_i, \ldots, a_{n+1})$$

represents a subdivision in two different ways of the polytope which is the convex hull of $(a_0, \ldots, a_{n+1})$. (See fig. for 3 cases in the plane.)

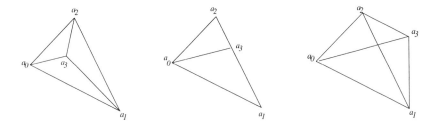

It may happen that one of the vertices of σ lies on a geodesic subspace spanned by some of the other. In that case some of the simplices in (2.7) are contained in a hyperplane and thus contributes zero in $\tilde{\mathcal{P}}(X)$. In general we define a filtration on $C_*(X)$

$$(2.8)$$
$$0 = C_*(X)^{-1} \subseteq C_*(X)^0 \subseteq \cdots \subseteq C_*(X)^p \subseteq \cdots \subseteq C_*(X)^n = C_*(X)$$

where $\sigma = (a_0, \ldots, a_k) \in C_*(X)^p$ if it is contained in a p-dimensional geodesic subspace. This filtration is called the *dimension* or *rank filtration* and n-simplices of filtration $n-1$ are called *flat*.

Now choose an orientation of X (usually the canonical one) and for a proper n-simplex $\sigma = (a_0, \ldots, a_n)$ put $\varepsilon_\sigma = +1$ or -1 depending on the orientation by the ordering of the vertices compared with the given orientation of X. Then there is an obvious map

(2.9) $$\varphi : C_n(X) \to \tilde{\mathcal{P}}(X)$$

defined for $\sigma = (a_0, \ldots, a_n)$ by

$$\varphi(\sigma) = \begin{cases} \varepsilon_\sigma \left[|\sigma|\right] & \text{if } \sigma \text{ is proper} \\ 0 & \text{otherwise,} \end{cases}$$

THEOREM 2.10. $\varphi : C_n(X) \to \tilde{\mathcal{P}}(X)$ *induces an isomorphism*

$$\varphi_* : H_n(C_*(X)/C_*(X)^{n-1}) \xrightarrow{\cong} \tilde{\mathcal{P}}(X)$$

Sketch proof. Since for $\sigma \in C_n(X)$ clearly $\partial\sigma \in C_*(X)^{n-1}$ we just have

$$H_n(C_*(X)/C_*(X)^{n-1}) = C_n(X)/\left[C_n(X)^{n-1} + \partial C_{n+1}(X)\right].$$

By definition φ maps $C_n(X)^{n-1}$ to zero. To see that φ also maps boundaries to zero we use a topological argument: Let $\Delta^k \subseteq \mathbb{R}^{k+1}$ be the standard simplex $\Delta^k = |(e_0, \ldots, e_k)|$ where $\{e_0, \ldots, e_k\}$ is the canonical basis. Then any k-simplex $\sigma = (a_0, \ldots, a_k)$ gives rise to a continuous map $f : \Delta^k \to |\sigma| \subseteq X$. Explicitly in the Euclidean case

$$f(t) = a(t) = \sum_{i=0}^{k} t_i a_i \quad \text{for } t = (t_0, \ldots, t_k) \in \Delta^k$$

whereas in the spherical respectively the hyperbolic cases

$$f(t) = \frac{a(t)}{\sqrt{\langle a(t), a(t) \rangle_+}} \quad \text{respectively} \quad f(t) = \frac{a(t)}{\sqrt{-\langle a(t), a(t) \rangle_-}}$$

where $\langle x, x \rangle_\varepsilon = \varepsilon x_0^2 + x_1^2 + \cdots + x_n^2, \varepsilon = \pm 1$, in the surrounding $\mathbb{R}^{n+1}$. In particular for any $(n+1)$-simplex $\sigma = (a_0, \ldots, a_{n+1})$ we get a continuous map

$$f : \partial\Delta^{n+1} \to \coprod_{i=0}^{n+1} |(a_0, \ldots, \hat{a}_i, \ldots, a_{n+1})| \subseteq X$$

of degree zero. If $\tau_i = (a_0, \ldots, \hat{a}_i, \ldots, a_{n+1})$ is proper then $\varepsilon_{\tau_i} = (-1)^i$ or $(-1)^{i+1}$ depending on f restricted to $|(e_0, \ldots, \hat{e}_i, \ldots, e_n)|$ being orientation preserving or reversing. Now we subdivide each proper simplex

$|\tau_i|$ by the hyperplanes supporting the other ones and we thus get a subdivision of the image of f such that in the sum

$$\sum_{\tau_i \text{ proper}} (-1)^i \varepsilon_{\tau_i} |\tau_i| = \varphi(\partial \sigma)$$

each piece occurs with multiplicity zero. Hence $\varphi(\partial\sigma) = 0$. It follows that φ induces a well-defined map φ_* which is clearly surjective. To show injectivity one can construct an inverse map using the simplicial approximation theorem. We refer to [**Dupont, 1982**] for details. For an alternative inductive argument see [**Morelli, 1993**].   □

Notice that the map φ is equivariant with respect to the natural action of $I(X)$ on $\tilde{\mathcal{P}}(X)$ and the "twisted" action on $C_n(X)$ given by

$$g(a_0, \ldots, a_n) = \det(g)(ga_0, \ldots, ga_n), \quad g \in I(X),$$

where $\det(g) = +1$ or -1 depending on g being orientation preserving or orientation reversing. In general for M any $I(X)$-module let M^t denote the module with the original action by g multiplied by $\det(g)$. With this notation we now conclude from proposition 2.1 and theorem 2.10:

COROLLARY 2.11. *Let $G \subseteq I(X)$ be a subgroup and fix an orientation of X. Then there are natural isomorphisms*

$$\mathcal{P}(X, G) \cong H_0\left(G, H_n(C_*(X)/C_*(X)^{n-1})^t\right)$$
$$\cong H_n\left(H_0(G, [C_*(X)/C_*(X)^{n-1}]^t)\right)$$

For later use, let us introduce the *polytope module*

(2.12) $\mathrm{Pt}(X) = H_n(C_*(X)/C_*(X)^{n-1})$

where $g \in I(X)$ acts by $g(a_0, \ldots, a_n) = (ga_0, \ldots, ga_n)$. Hence as an $I(X)$-module

(2.13) $\tilde{\mathcal{P}}(X) \cong \mathrm{Pt}(X)^t$

and we reformulate corollary 2.11 as

(2.14) $\mathcal{P}(X, G) \cong H_0(G, \mathrm{Pt}(X)^t).$

In the spherical case it is also useful to consider the chain complex $\bar{C}_*(S^n)$ where we have dropped the restriction that the vertices $(a_0, \ldots, a_k)$

of a k-simplex should lie on a hemi-sphere. Then we introduce the *Steinberg module*

$$(2.15) \qquad \mathrm{St}(S^n) = H_n(\bar{C}_*(S^n)/\bar{C}_*(S^n)^{n-1})$$

and in the Euclidean and hyperbolic cases we just put

$$(2.16) \qquad \mathrm{St}(E^n) = \mathrm{Pt}(E^n) \quad , \quad \mathrm{St}(\mathcal{H}^n) = \mathrm{Pt}(\mathcal{H}^n).$$

In the next chapter we shall give a different description ot these $I(X)$-modules.

Note. Cohomological arguments in connection with s.c. appeared already in [**Jessen-Karpf-Thorup, 1968**]. A systematic use of group cohomology was made in [**Sah, 1979**] inspired by remarks of D. Sullivan. The interpretation of the s.c. group as a homology group (corollary 2.11) comes from [**Dupont, 1982**], but the geometric content of theorem 2.10 that $\tilde{\mathcal{P}}(X)$ is generated by simplices modulo the elementary subdivisions represented by (2.7) was known previously to B. Jessen and A. Thorup (unpublished note by Thorup). In fact they proved that it suffices to use only "simple subdivisions":

$$|(a_0, \dots, a_n)| = |(a_0, \dots, a_{n-1}, a_{n+1})| \amalg |(a_0, \dots, a_{n-2}, a_n, a_{n+1})|$$

where a_{n+1} lies on the edge (a_{n-1}, a_n). Note by the way that the s.c. group which was introduced by Jessen in 1941 (c.f. [**Jessen, 1941**]) is really the K-group in the sense of algebraic K-theory for the category of polytopes.

CHAPTER 3

Homology of flag complexes

In this chapter we shall investigate the Steinberg and polytope modules further and in particular we shall establish some exact sequences relating these modules in different dimensions which are useful for inductive arguments later. First let us give another interpretation of the Steinberg module as the homology of a complex of flags of *subspaces*. The idea is that a geometric n-simplex in X is determined by its supporting hyperplanes (i.e. the subspaces spanned by its faces of codimension 1) at least in the case of $X = E^n$ or $\mathcal{H}^n$. By a *subspace* of X we mean a geodesic subspace U of a certain dimension p such that U is a model for the geometry in that dimension. In E^n a subspace is an *affine subspace*, in $\mathcal{H}^n$ respectively S^n it is a *hyperbolic* respectively *spherical* subspace. In our models for $\mathcal{H}^n$ and S^n these are just intersections with linear subspaces of the surrounding $\mathbb{R}^{n+1}$. Note in particular that a 0-dimensional spherical subspace consists of 2 antipodal points. If $A \subseteq X$ is any set, then by span A we mean the smallest subspace (in the above sense) containing A. E.g. in the spherical case span$\{a\} = \{\pm a\}$ for $a \in S^n$.

Next recall that a *simplicial set* is a sequence of sets of *simplices* $S = \{S_p; p = 0, 1, 2, \ldots\}$ together with *face operators* $\varepsilon_i : S_p \to S_{p-1}, i = 0, \ldots, p$, and *degeneracy operators* $\eta_j : S_p \to S_{p+1}, j = 0, \ldots, p$, satisfying the identities.

(3.1) $\qquad \varepsilon_i \varepsilon_j = \varepsilon_{j-1} \varepsilon_i, \quad i < j; \quad \eta_i \eta_j = \eta_{j+1} \eta_i, \quad i \leq j,$

and

(3.2) $\qquad \varepsilon_i \eta_j = \begin{cases} \eta_{j-1} \varepsilon_i, & i < j \\ \text{id}, & i = j, \quad i = j+1 \\ \eta_j \varepsilon_{i-1}, & i > j+1. \end{cases}$

Our main example is the following:

> DEFINITION 3.3.   a) The *Tits complex* $\mathcal{T}(X)$ is the simplicial set
> where a p-simplex Φ is a *flag* $\Phi = (U_0 \supseteq \cdots \supseteq U_p)$ of *proper* (i.e.
> $U_i \neq X, \emptyset$) subspaces with face and degeneracy operators given
> by
>
> $$\varepsilon_i(U_0 \supseteq \cdots \supseteq U_p) = (U_0 \supseteq \cdots \supseteq \hat{U}_i \supseteq \ldots U_p)$$
> $$\eta_j(U_0 \supseteq \cdots \supseteq U_p) = (U_0 \supseteq \cdots \supseteq U_j \supseteq U_j \supseteq \cdots \supseteq U_p)$$
>
> b) A flag $\Phi = (U_0 \supseteq \cdots \supseteq U_p)$ is called *strict* if U_i has co-dimension
> $i + 1, i = 0, \ldots, p$, and $p + 1$ is called the *length* of Φ.

For a simplicial set S the (reduced) integral homology groups are the
homology groups of the (augmented) chain complex $C_*(S)$, where $C_p(S)$
is the free abelian group on S_p and the boundary ∂ (and augmentation
$\varepsilon : C_0(S) \to \mathbb{Z}$) is given by

$$\partial(\sigma) = \sum_{i=0}^{p}(-1)^i\varepsilon_i(\sigma) \quad , \quad \sigma \in C_p(S) \quad , \quad p = 1, 2, \ldots,$$
(3.4)
$$\varepsilon(\sigma) = 1 \quad , \quad \sigma \in C_0(S).$$

With this notation we have:

THEOREM 3.5. *Let $\tilde{H}_*(\mathcal{T}(X))$ denote the reduced integral homology
of $\mathcal{T}(X)$. Then*

a) $\tilde{H}_i(\mathcal{T}(X)) = 0$ *if $n \geq 1$ and $i \neq n - 1$*
b) *There are natural isomorphisms*

$$\mathrm{St}(X^0) \cong \mathbb{Z} \text{ and } \mathrm{St}(X^n) \cong \tilde{H}_{n-1}(\mathcal{T}(X)), \quad n \geq 1.$$

c) *In b) $\mathrm{St}(X^n)$ is identified with the following subgroup of $\bigoplus_{\Phi} \mathbb{Z}(\Phi)$,
where Φ runs through all strict flags of length n: Here $x = \sum_{\Phi} x_\Phi$
lies in $\mathrm{St}(X)$ if and only if for all $i = 0, \ldots, n - 1$ and all fixed
$(U_0 \supseteq \cdots \supseteq U_{i-1} \supseteq U_{i+1} \supseteq \cdots \supseteq U_{n-1})$ we have*

$$\sum_{U_i} x_{(U_0 \supseteq \cdots \supseteq U_{n-1})} = 0.$$

For the proof we consider for each subspace $U \subseteq X$ the chain complex $\bar{C}_*(U)$ which in degree q is the free abelian group with generators all $(q+1)$-tuples of points in U and with the boundary given by formula (2.6) (so that $\bar{C}_*(U) = C_*(U)$ in the Euclidean and hyperbolic case).

LEMMA 3.6. $H_q(\bar{C}_*(U)) = \begin{cases} \mathbb{Z}, & q = 0 \\ 0 & otherwise. \end{cases}$

Proof: We augment $\bar{C}_*(U)$ to $\mathbb{Z}$ by $\varepsilon : \bar{C}_0(U) \to \mathbb{Z}$ given by $\varepsilon(a_0) = 1$ and observe that the sequence

$$\cdots \xrightarrow{\partial} \bar{C}_q(U) \xrightarrow{\partial} \cdots \xrightarrow{\partial} \bar{C}_0(U) \xrightarrow{\varepsilon} \mathbb{Z} \to 0$$

is exact. In fact if we choose a base point $u_0 \in U$ we get a chain contraction $s_q : \bar{C}_q(U) \to \bar{C}_{q+1}(U)$ defined by

$$s_q(a_0, \ldots, a_q) = (u_0, a_0, \ldots, a_q), \quad q \geq 0, \text{ and } s_{-1}(1) = (u_0).$$

Then it is easy to check that

$$\partial s + s \partial = \mathrm{id}$$

which proves the exactness.   □

Proof of theorem 3.5 Consider the bicomplex (cf. Appendix A)

$$(3.7) \qquad A_{p,*} = \bigoplus_{(U_0 \supseteq \cdots \supseteq U_p)} \bar{C}_*(U_p), \quad p \geq 0,$$

augmented to $\bar{C}_*(X)^{n-1}$ by the obvious inclusion maps $\bar{C}_*(U_0) \subseteq \bar{C}_*(X)$. Here the "vertical" differential is $\partial'' = (-1)^p \partial$ in $\bar{C}_*(U_p)$ and the "horizontal" differential is $\partial' = \sum_{i=0}^p (-1)^i \varepsilon_{i*}$. Then the sequence

$$0 \leftarrow \bar{C}_*(X)^{n-1} \leftarrow A_{0,*} \xleftarrow{\partial'} A_{1,*} \xleftarrow{\partial'} \cdots$$

is exact. In fact we define $s_p : A_{p,*} \to A_{p+1,*}$ for a simplex $\sigma \subseteq U_p$ by

$$s_p(\sigma_{(U_0 \supseteq \cdots \supseteq U_p)}) = (-1)^{p+1} \sigma_{(U_0 \supseteq \cdots \supseteq U_p \supseteq U_\sigma)}, \quad p = -1, 0, 1, \ldots,$$

where $U_\sigma = \mathrm{span}(\sigma)$, and it is straightforward to check that

$$\partial' s + s \partial' = \mathrm{id}$$

It follows that the homology of the total complex $(A_*, \partial' + \partial'')$, where $A_k = \bigoplus_{p+q=k} A_{p,q}$, is given by

$$H(A_*) \cong H(\bar{C}_*(X)^{n-1}).$$

By the exact sequence

$$0 \to \bar{C}_*(X)^{n-1} \to \bar{C}_*(X) \to \bar{C}_*(X)/\bar{C}_*(X)^{n-1} \to 0$$

and lemma 3.6 we also have

$$\tilde{H}_{k-1}(\bar{C}_*(X)^{n-1}) \cong H_k(\bar{C}_*(X)/\bar{C}_*(X)^{n-1}).$$

On the other hand lemma 3.6 also gives an exact sequence

$$0 \leftarrow C_p(\mathcal{T}(X)) \leftarrow A_{p,0} \overset{\partial''}{\leftarrow} A_{p,1} \overset{\partial''}{\leftarrow} A_{p,2} \overset{\partial''}{\leftarrow} \cdots$$

where

$$C_p(\mathcal{T}(X)) = \underset{\Phi=(U_0 \supseteq \cdots \supseteq U_p)}{\bigoplus} \mathbb{Z}\Phi$$

is the p-th chain group for $\mathcal{T}(X)$. Also ∂'' clearly induces the boundary map on $C_*(\mathcal{T}(X))$ so that we obtain

$$H(A_*) \cong H_*(\mathcal{T}(Z)).$$

It follows that we have an isomorphism

$$H_k(\bar{C}_*(X)/\bar{C}_*(X)^{n-1}) \cong \tilde{H}_{k-1}(\mathcal{T}(X)), \quad k \geq 1.$$

Here the left hand side is clearly zero for $k < n$ whereas the right hand side is zero for $k > n$ since $\mathcal{T}(X)$ contains only degenerate simplices in these dimensions. This proves a) and b). For c) we observe that $H_*(\mathcal{T}(X))$ is computed by the chain complex $C_*^N(\mathcal{T}(X))$ with only *non-degenerate* generators, i.e. in degree $n-1$ we have only *strict* flags, and hence

$$H_{n-1}(\mathcal{T}(X)) = \ker(\partial : C_{n-1}^N(\mathcal{T}(X)) \to C_{n-2}^N(\mathcal{T}(X))).$$

But here $\partial(x) = 0$ if and only if $\varepsilon_i(x) = 0$, $i = 0, \cdots, n-1$, since $\varepsilon_i(x)$ and $\varepsilon_j(x)$ involve flags of different types for $i \neq j$. □

Remark: The isomorphism in theorem 3.5 can be explicitly described as induced by the map

$$h : \bar{C}_n(X) \to C_{n-1}(\mathcal{T}(X))$$

given by

$$(3.8) \qquad h(a_0, \ldots, a_n) = \sum_{\pi} \text{sign}(\pi)(U_0^\pi \supseteq \cdots \supseteq U_{n-1}^\pi)$$

where π runs through all permutations of $\{0, \ldots, n\}$ and

$$U_j^\pi = \text{span}\{a_{\pi(j+1)}, \ldots, a_{\pi(n)}\}.$$

Note that this map is equivariant with respect to the natural action by $I(X)$.

Example:

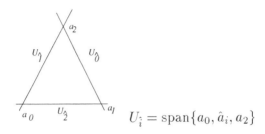

$$U_{\hat{i}} = \text{span}\{a_0, \hat{a}_i, a_2\}$$

$$h(a_0, a_1, a_2) = (U_{\hat{0}} \supset a_2) - (U_{\hat{0}} \supset a_1) + (U_{\hat{1}} \supset a_0) - (U_{\hat{1}} \supset a_2) +$$
$$+ (U_{\hat{2}} \supset a_1) - (U_{\hat{2}} \supset a_0)$$

The corresponding theorem in the case of the polytope module for $X = S^n$ is a little more complicated since the topology of the sphere interferes. Thus a geodesic subspace $U \subseteq X$ of dimension p is a p-sphere and if we consider the chain complex $C_*(U) \subseteq C_*(X)$ as in chapter 2 where all vertices of a simplex in U lie on an open hemisphere then the analogue of lemma 3.6 is

$$(3.9) \qquad H_q(C_*(U)) = \begin{cases} \mathbb{Z} & q = 0 \text{ or } p \\ 0 & \text{otherwise.} \end{cases}$$

In fact, as mentioned in the proof of theorem 2.10 a simplex $\sigma = (a_0, \ldots, a_q)$ in U gives rise to a singular q-simplex $f_\sigma : \Delta^q \to U$ and one shows (cf. proposition A.21) that the inclusion $C_*(U) \to C_*^{\text{sing}}(U)$ into the singular chain complex induces an isomorphism in homology. Let us write $O(U) = H_p(U) \cong \mathbb{Z}$ and call it the *orientation module* for U since an isometry g of U induces multiplication by ± 1 depending on g being orientation preserving or not. Thus $O(X) = \mathbb{Z}^t$ in our previous notation. Also we put $O(\emptyset) = \mathbb{Z}$. Finally for $U \subseteq X$ a p-dimensional subspace let $U^\perp$ denote the subspace perpendicular to U. We can now state the analogue of theorem 3.5 for the polytope module. We refer to [**Dupont, 1982**] for the proof.

THEOREM 3.10. *Let* $X = S^n, n \geq 0$. *Then*

i) $H_q(C_*(X)/C_*(X)^{n-1}) = 0$ *for* $q \neq n$.
ii) *There is a filtration of* $I(X)$-*modules*

$$0 \subseteq O(X) = F_{-1} \subseteq F_0 \subseteq \cdots \subseteq F_p \subseteq \cdots \subseteq F_n = \mathrm{Pt}(X)$$

and natural isomorphisms

$$F_p/F_{p-1} \cong \bigoplus_{U^p} \mathrm{St}(U^p) \otimes O(U^{p\perp}), \quad p = 0, 1, 2, \ldots, n.$$

where U^p *runs through all* p-*dimensional subspaces of* X.
iii) *In particular*

$$\mathrm{Pt}(X)/F_{n-1} \cong \mathrm{St}(X).$$

Remark: Geometrically F_p is generated by chains of the form a join of a p-simplex in U^p with a triangulation of the sphere $U^{p\perp}$, where U^p and $U^{p\perp}$ are perpendicular subspaces of dimension p respectively $n - p - 1$.

The Steinberg and polytope modules give rise to some useful exact sequences relating these to the corresponding ones in lower dimensions. For the Steinberg modules we have in all 3 geometries:

THEOREM 3.11. *Let* $\dim X = n$. *Then there is a natural exact sequence of* $I(X)$-*modules*

$$0 \to \mathrm{St}(X) \xrightarrow{d} \bigoplus_{U^{n-1}} \mathrm{St}(U^{n-1}) \xrightarrow{d} \bigoplus_{U^{n-2}} \mathrm{St}(U^{n-2}) \to \cdots$$

$$\cdots \xrightarrow{d} \bigoplus_{U^0} \mathrm{St}(U^0) \xrightarrow{\varepsilon} \mathbb{Z} \to 0$$

where U^p *runs though all* p-*dimensional subspaces of* X.

Remark: Here $d : \mathrm{St}(U^p) \to \bigoplus_{U^{p-1}} \mathrm{St}(U^{p-1})$ is induced by sending a strict flag $(U_0 \supset \cdots \supset U_{p-1})$ for U^p to the flag $(U_1 \supset \cdots \supset U_{p-1})$ for $U^{p-1} = U_0$. Also $\mathrm{St}(U^0) = \mathbb{Z}(U^0)$ and ε is the augmentation $\varepsilon(U^0) = 1$.

Proof: First notice that by the proof of theorem 3.5 we have for U a p-dimensional subspace

$$H_i(\bar{C}_*(U)/\bar{C}_*(U)^{p-1}) = \begin{cases} \text{St}(U) & i = p \\ 0 & \text{otherwise.} \end{cases}$$

It follows that the spectral sequence for the filtration $F_p\bar{C}_*(X) = \bar{C}_*(X)^p$ has

$$E_{p,*}^0 = F_p\bar{C}_*(X)/F_{p-1}\bar{C}_*(X) \cong \bigoplus_{U^p} \bar{C}_*(U^p)/\bar{C}_*(U^p)^{p-1}$$

and hence

$$E_{p,q}^1 = \bigoplus_{U^p} H_{p+q}(\bar{C}_*(U^p)/\bar{C}_*(U^p)^{p-1}) = \begin{cases} \bigoplus_{U^p} \text{St}(U^p), & q = 0 \\ 0, & q \neq 0 \end{cases}$$

It follows that $E_{p,q}^2 = E_{p,q}^\infty$. But by lemma 3.6 this is zero except for $p = q = 0$ where it is $\mathbb{Z}$. Hence we get an exact sequence

$$0 \to E_{n,0}^1 \xrightarrow{d^1} E_{n-1,0}^1 \xrightarrow{d^1} \cdots \xrightarrow{d^1} E_{0,0}^1 \xrightarrow{\varepsilon} \mathbb{Z} \to 0$$

which is just the required sequence. □

For the polytope module in the spherical case we get in a similar way using theorem 3.10:

THEOREM 3.12. *Let* $X = S^n, n \geq 0$. *Then there is an exact sequence of* $I(X)$-*modules*

$$0 \to \mathbb{Z}^t \xrightarrow{i} \text{Pt}(X) \xrightarrow{d} \bigoplus_{U^{n-1}} \text{Pt}(U^{n-1}) \xrightarrow{d} \bigoplus_{U^{n-2}} \text{Pt}(U^{n-2}) \to \cdots$$

$$\cdots \xrightarrow{d} \bigoplus_{U^0} \text{Pt}(U^0) \xrightarrow{\varepsilon} \mathbb{Z} \to 0$$

where U^p *runs through all* p-*dimensional subspaces of* X.

Remark 1: The inclusion $i : \mathbb{Z}^t \to \text{Pt}(X)$ is just the natural map

$$\mathbb{Z}^t \cong H_n(C_*(S^n)) \to H_n(C_*(S^n)/C_*(S^n)^{n-1}) = \text{Pt}(S^n).$$

The generator on the left is sent to the polytope consisting of the whole sphere (suitably subdivided).

Remark 2: For $X = S^0$ the sequence reduces to the exact sequence of $O(1) = \{\pm 1\}$-modules:

$$0 \to \mathbb{Z}^t \to \text{Pt}(S^0) \to \mathbb{Z} \to 0$$

Finally there is in the spherical case an exact sequence combining the Steinberg and polytope modules. For $U \subseteq X = S^n$ a subspace of dimension $p < n$ and $\{\pm e\}$ a pair of antipodal points perpendicular to U there is a *suspension homomorphism*

$$\Sigma_U : \text{Pt}(U)^t \to \text{Pt}(V)^t, \quad V = \text{span}(U \cup \{\pm e\}),$$

given by the formula

$$\Sigma_U((a_0, \ldots, a_p) \otimes \sigma) = [(e, a_0, \ldots, a_p) - (-e, a_0, \ldots, a_p)] \otimes (e * \sigma).$$

Here we use the notation $\text{Pt}(U)^t = \text{Pt}(U) \otimes O(U)$ and for $\sigma \in O(U) \cong \mathbb{Z}^t$ a generator we let $e * \sigma$ be the generator of $O(V)$ determined by e and σ. This suspension extends to a map

$$\Sigma_U : \text{Pt}(U)^t \otimes \text{St}(U^\perp) \to \bigoplus_{\dim V = p+1} \text{Pt}(V)^t \otimes \text{St}(V^\perp)$$

in a straightforward manner (notice that a strict flag occuring in $\text{St}(U^\perp)$ ends with a pair of antipodal points perpendicular to U). In particular for $U = \emptyset$ we have

$$\Sigma : \text{St}(X) \to \bigoplus_{U^0} \text{Pt}(U^0) \otimes (U^{0\perp})$$

and together with the map

$$h : \text{Pt}(X) \to \text{St}(X)$$

given by the formula in (3.8) we obtain the following theorem wich we state without proof:

THEOREM 3.13. *For* $X = S^n, n \geq 0$, *there is an exact sequence of* $I(X)$-*modules*

$$0 \to \mathrm{St}(X) \overset{\Sigma}{\to} \bigoplus_{U^0} \mathrm{Pt}(U^0)^t \otimes \mathrm{St}(U^{0\perp}) \overset{\Sigma}{\to} \ldots$$

$$\ldots \to \bigoplus_{U^p} \mathrm{Pt}(U^p)^t \otimes \mathrm{St}(U^{p\perp}) \overset{\Sigma}{\to} \ldots$$

$$\ldots \overset{\Sigma}{\to} \bigoplus_{U^{n-1}} \mathrm{Pt}(U^{n-1})^t \otimes \mathrm{St}(U^{n-1\perp}) \overset{\Sigma}{\to} \mathrm{Pt}(X)^t \overset{h}{\to} \mathrm{St}(X)^t \to 0.$$

Note: Exact sequences like the one in theorem 3.11 are called *Lusztig exact sequences* in analogy to those constructed in [**Lusztig, 1974**]. However, similar sequences were known by J. Karpf (unpublished Msc. thesis, Copenhagen University,1969).

CHAPTER 4

Translational scissors congruences

In the Euclidean case $X = E^n$, the group $T(n)$ of translations is a normal subgroup of the group of all isometries with the orthogonal group $O(n)$ as quotient group. Hence by proposition 2.1

$$(4.1) \qquad \mathcal{P}(E^n) = H_0(O(n), \mathcal{P}(E^n, T(n))).$$

In this chapter we shall study $\mathcal{P}(E^n, T(n))$ in more detail, and in particular we shall find necessary and sufficient conditions for translational s.c. (the Hadwiger invariants). More generally let us consider a field F of characteristic 0 and V a vector space of dimension n over F. The group of *translations* of V is just the additive group of V acting on V by $t_v(x) = x + v$, $\quad x, v \in V$. Again $C_*(V)$ denote the chain complex over $\mathbb{Z}$ of tuples of elements of V and in view of corollary 2.11 we want to calculate the *translational scissors congruence group* defined by

$$(4.2) \quad \mathcal{P}_T(V) = H_n\left(\frac{C_*(V)}{V} \Big/ \frac{C_*(V)^{n-1}}{V}\right), \quad \frac{C_*(V)}{V} = H_0(V, C_*(V)).$$

Now $C_*(V)$ is just the homogeneous "bar complex" for the *additive* group of V. Let us recall a few facts about *group homology* :

For $(G, \cdot)$ an arbitrary group and $\mathbb{Z}[G]$ the group ring the *homogeneous bar complex* is the complex $C_*(G)$ with generators in degree q any $(q+1)$-tuple $(a_0, \ldots, a_q)$ and with the usual boundary map

$$\partial(a_0, \ldots, a_q) = \sum_{i=1}^{q} (-1)^i (a_0, \ldots, \hat{a}_i, \ldots, a_q)$$

This is a complex of left $\mathbb{Z}[G]$-modules by the action

$$g(a_0, \ldots, a_q) = (ga_0, \ldots, ga_q), \quad g \in G,$$

and it is a resolution of the trivial G-module $\mathbb{Z}$ via the augmentation $\varepsilon : C_0(G) \to \mathbb{Z}$, $\varepsilon(a_0) = 1$. Sometimes we shall use the *inhomogeneous* notation: Let $B_*(G)$ be the complex with generators in degree q of the form $g_0[g_1 \mid \cdots \mid g_q]$, $g_i \in G$, and boundary map given by

$$\partial(g_0[g_1 \mid \cdots \mid g_q]) = g_0g_1[g_2 \mid \cdots \mid g_q] + \sum_{i=1}^{q-1}(-1)^i g_0[g_1 \mid \cdots \mid g_ig_{i+1} \mid \ldots g_q]$$
$$+ (-1)^q g_0[g_1 \mid \ldots g_{q-1}].$$

Here the G-action is given by

$$g(g_0[g_1 \mid \cdots \mid g_q]) = gg_0[g_1 \mid \cdots \mid g_{q-1}]$$

and the augmentation ε to $\mathbb{Z}$ by $\varepsilon(g_0[\cdot]) = 1$. Actually $C_*(G)$ and $B_*(G)$ are isomorphic via the bijection $(a_0, \ldots, a_q) \leftrightarrow g_0[g_1 \mid \cdots \mid g_q]$ where

$$a_0 = g_0, \; a_1 = g_0g_1, \quad \cdots \quad , a_q = g_0 \cdots g_q.$$

For a *right* G-module N the homology of G is by definition $H_*(G, N) = H(N \otimes_{\mathbb{Z}[G]} B_*(G))$, so for M a *left* G module $H_*(G, M) = H_*(G, M^{\mathrm{Op}})$, where M^{Op} has the right G-action $xg = g^{-1}x$, $x \in M$, $g \in G$. Equivalently it is the homology of the complex $\bar{B}_*G \otimes M$ with $\bar{B}_0G = \mathbb{Z}$ and for $q > 0$ generated by symbols $[g_1 \mid \cdots \mid g_q] \otimes x$, $g_1, \ldots, g_q \in G$, $x \in M$, and with boundary map

$$\partial([g_1 \mid \cdots \mid g_q]) \otimes x) = (g_2 \mid \cdots \mid g_q] \otimes g_1^{-1}x +$$
(4.3)
$$+ \sum_{i=1}^{q-1}(-1)^i[g_1 \mid \cdots \mid g_ig_{i+1} \mid \cdots \mid g_q] \otimes x$$
$$+ (-1)^q[g_1 \mid \cdots \mid g_{q-1}] \otimes x.$$

Notice that

(4.4) $\qquad H_0(G, M) = M/\{x - gx \mid x \in M, g \in G\} = G \backslash M = \dfrac{M}{G}.$

For $M = \mathbb{Z}$ the trivial G-module we write

$$H_*(G) = H_*(G, \mathbb{Z}) = H_*(\bar{B}_*(G)) = H_*(G \backslash C_*(G))$$

For G and G' two groups we shall need the *Eilenberg-Zilber map* of $G \times G'$-modules:

$$EZ : B_*(G) \otimes B_*(G') \to B_*(G \times G')$$

given by

$$EZ(g[g_1 \mid \cdots \mid g_p] \otimes g'[g'_{p+1} \mid \cdots \mid g'_{p+q}])$$
$$= \sum_\sigma \text{sign } \sigma \, (g, g')[h_{\sigma 1} \mid \cdots \mid h_{\sigma(p+q)}]$$

where $(h_1 \ldots h_{p+q}) = ((g_1, 1) \ldots, (g_p, 1), (1, g'_{p+1}), \ldots (1, g'_{p+q}))$ and σ runs through all (p, q) shuffles of $1, \ldots, p + q$. For M resp. M' G - resp. G'-modules EZ induces an $\times$-product

$$\times : H_*(G, M) \otimes H_*(G', M') \overset{EZ_*}{\to} H_*(G \times G', M \otimes M'),$$

and by the Eilenberg-Zilber and Künneth theorems we get an exact sequence

(4.5)
$$0 \to H_*(G, M) \otimes H_*(G', M') \overset{\times}{\to} H_*(G \times G', M \times M') \to$$
$$\to \left[\text{Tor}(H_*(G, M), H_*(G', M)) \right]_{-1} \to 0$$

where the subscript -1 indicates that in degree k the indices add up to $k - 1$.

Now suppose $(A, +)$ is an abelian group. Then $+ : A \times A \to A$ is a homomorphism and we define the *Pontrjagin product* as the composite

(4.6) $\wedge : H_i(A) \otimes H_j(A) \overset{\times}{\to} H_{i+j}(A \times A) \overset{+_*}{\to} H_{i+j}(A).$

This makes $(H_*(A), \wedge)$ into a graded ring.

PROPOSITION 4.7. *Let A be an abelian group. Then*

  i) $H_0(A) = \mathbb{Z}$ *and there is a natural isomorphism* $A \cong H_1(A)$.
  ii) *If A is torsion free then the $\wedge$-product defines a natural isomorphism* $\wedge^k_{\mathbb{Z}}(A) \cong \wedge^k_{\mathbb{Z}}(H_1(A)) \overset{\wedge^k}{\to} H_k(A), \quad k = 0, 1, 2, \ldots.$
  iii) *If A is a divisible group then $A \cong A/T \oplus T$, where T is the torsion subgroup of A, and we have an isomorphism*
$$H_k(A) \cong \wedge^k_{\mathbb{Q}}(A/T) \oplus H_k(T)$$

Proof: i) The first statement is obvious from (4.4), and for the second we notice that

$$H_1(A) = \bar{B}_1(A)/\partial \bar{B}_2(A) = \mathbb{Z}[A]/\{[a] + [b] - [a + b] \mid a, b \in A\},$$

so that the natural map $A \to H_1(A)$ sending a to $[a]$ is an isomorphism.

ii) Since both sides of the equation commutes with direct limits it is enough to consider A a finitely generated torsion free abelian group, that is, $A \cong \mathbb{Z} \oplus \cdots \oplus \mathbb{Z}$ in which case it follows from the Künneth isomorphism (4.5).

iii) This similarly follows from (4.5) and the observation that

$$\wedge_{\mathbb{Z}}^k(A/T) \cong \wedge_{\mathbb{Q}}^k(A/T)$$

since A/T is a uniquely divisible group, i.e. a rational vector space.

Remark: Notice that the isomorphism in ii) is explicitly given by sending $u_1 \wedge \cdots \wedge u_k \in \wedge_{\mathbb{Z}}^k(A)$ to

(4.8) $$u_1 \wedge \cdots \wedge u_k = \sum_\sigma \operatorname{sign} \sigma \, [u_{\sigma 1} \mid \cdots \mid u_{\sigma k}]$$

where σ runs through all permutations of $1, \ldots, k$.

We now return to the computation of $\mathcal{P}_T(V)$ for V an n-dimensional vector space over a field F of characteristic 0. First notice that by proposition 4.7 we have a natural isomorphism for $U \subseteq V$ any linear subspace:

(4.9) $$\wedge_{\mathbb{Q}}^k(U) \cong H_k\left(\frac{C_*(U)}{U}\right), \quad k = 0, 1, \ldots$$

where we have used the isomorphism $C_*(U) \cong B(U)$ given by

$$(x_0, \ldots, x_k) \leftrightarrow x_0[u_1 \mid \cdots \mid u_k]$$

with

$$x_1 = x_0 + u_1, \, x_2 = x_0 + u_1 + u_2, \ldots, x_k = x_0 + u_1 + u_2 + \cdots + u_k.$$

Geometrically the element given by (4.8) corresponds to the k-fold cube $[u_1] \wedge \cdots \wedge [u_k]$ where $[u_i] = (0, u_i) \in C_1(U)$.

For the computation of the homology group in (4.2) we now proceed as in chapter 3. We let $\mathcal{T}(V)$ be the simplicial set where a p-simplex Φ is a flag $\Phi = (U_0 \supsetneq \cdots \supsetneq U_p)$ of *proper* linear subspaces of V, i.e. different from 0 and V. Notice that for $V = \mathbb{R}^n$ there is a natural bijection $\mathcal{T}(V) \cong \mathcal{T}(S^{n-1})$, in the notation of chapter 3, given by intersecting the subspaces with S^{n-1}. We now consider the bicomplex $A_{p,q}$ for $p, q \geq -1$ given by

$$A_{p,*} = \begin{cases} V \backslash \tilde{C}_*(V), & p = -1 \\ \bigoplus_{\Phi = (U_0 \supseteq \cdots \supseteq U_p)} U_p \backslash \tilde{C}_*(U_p), & p = 0, 1, 2, \ldots \end{cases}$$

where $\tilde{C}_*(U), U \subseteq V$, is the chain complex $C_*(U)$ augmented in the usual way to $\tilde{C}_{-1}(U) = \mathbb{Z}$. Also the boundary maps ∂' and ∂'' are induced by the corresponding maps in the proof of theorem 3.5. Again the homology with respect to ∂' vanishes for $p \geq 0$ and we conclude that the total complex has homology

$$H_k(A_*) \cong H_{k+1}(V \backslash C_*(V) / V \backslash C_*(V)^{n-1}), \quad k = -1, 0, 1, 2, \ldots$$

Thus we get a spectral sequence converging to $H(A_*)$ with

$$E^1_{p,*} \cong \begin{cases} \tilde{H}_*(V) & p = -1 \\ \bigoplus_{U_0 \supseteq \cdots \supseteq U_p} \tilde{H}_*(U_p) & p = 0, 1, 2, \ldots \end{cases}$$

where, by (4.9) above $\tilde{H}_q(U_p) = \wedge^q_{\mathbb{Q}}(U_p)$, $q > 0$, and zero for $q = 0, -1$. We therefore get

$$E^2_{p,q} \cong \begin{cases} \tilde{H}_p(\mathcal{T}(V), \wedge^q_{\mathbb{Q}}(\mathfrak{g})) & q > 0 \\ 0 & q = 0, -1 \end{cases}$$

where $\wedge^q_{\mathbb{Q}}(\mathfrak{g})$ denotes the local coefficient system on $\mathcal{T}(V)$ given by

$$\wedge^q_{\mathbb{Q}}(\mathfrak{g})_{(U_0 \supseteq \cdots \supseteq U_p)} = \wedge^q_{\mathbb{Q}}(U_p)$$

and where the chain groups are augmented to $\tilde{C}_{-1}(\mathcal{T}(V), \wedge^q_{\mathbb{Q}}(\mathfrak{g})) = \wedge^q_{\mathbb{Q}}(V)$.

However, the spectral sequence degenerates from the E^2-level as we shall now see: Choose a positive integer $a > 1$ and let $\mu_a : V \to V$ be multiplication by a. The translation $t_v : V \to V$ by $v \in V$ clearly satisfies

$$\mu_a \circ t_v = t_{av} \circ \mu_a$$

and it follows that μ_a induces an endomorphism of the double complex $A_{*,*}$ and hence of the spectral sequence commuting with the differentials. However it is easy to see that the induced map on $E^1_{p,q}$ is just multiplication by a^q and since $d^r : E^r_{p,q} \to E^r_{p-r,q+r-1}$, it follows that $d^r = 0$ for $r \geq 2$. We have thus proved

THEOREM 4.10.   a) $H(V\backslash C_*(V)/V\backslash C_*(V)^{n-1})$ and in particular $\mathcal{P}_T(V)$ are vector spaces over $\mathbb{Q}$.

b) There is a splitting into eigenspaces for the endomorphism μ_a
with eigenvalues a^q, $q = 1, 2, \ldots, n$:

$$\mathcal{P}_T(V) \cong \bigoplus_{q=1}^{n} \tilde{H}_{n-q-1}(\mathcal{T}(V), \wedge_{\mathbb{Q}}^q(\mathfrak{g})),$$

and the splitting is independent of choice of integer $a > 1$.

c) $\tilde{H}_p(\mathcal{T}(V), \wedge_{\mathbb{Q}}^q(\mathfrak{g})) = 0$ for $p < n - q - 1$, $q > 0$.

Remark: The spectral sequence gives rise to a filtration $F_p\mathcal{P}_T(V)$, $p = -1, 0, 1, \ldots, n-2$, such that $F_p/F_{p-1} \cong \tilde{H}_p(\mathcal{T}(V), \wedge_{\mathbb{Q}}^q(\mathfrak{g}))$ with $p + q = n - 1$ which corresponds to the a^q-eigenspace for μ_a. It follows that any element $x \in F_p\mathcal{P}_T(V)$ has a unique decomposition

(4.11) $x = x_q + x_{q+1} + \ldots$

such that x_i has "weight i", that is, is in the a^ith eigenspace. We will now get a geometric interpretation for this filtration.

Definition:

a) A k-fold *prism* $x \in \mathcal{P}_T(V)$ is an element of the form

$$x = [v_1 \mid \cdots \mid v_{i_1}] \wedge [v_{i_1+1} \mid \cdots \mid v_{i_1+i_2}] \wedge \cdots \wedge [v_{i_1+\cdots+i_{k-1}+1} \mid \cdots \mid v_{i_1+\cdots+i_k}]$$

where $i_1 + \cdots + i_k = n$ and $v_1, \ldots, v_n \in V$.

b) Let $\mathcal{Z}_k(V) \subseteq \mathcal{P}_T(V)$ be the subgroup generated by all k-fold prisms.

PROPOSITION 4.12. $\mathcal{Z}_k(V)$ is the sum of eigenspaces of weight $\geq k$.
That is, $\mathcal{Z}_k(V) = F_{n-k-1}\mathcal{P}_T(V)$.

Proof: Every element has weight at least 1 so at least $\mathcal{Z}_k(V) \subseteq F_{n-k-1}$. For the other inclusion it suffices by induction to prove that every element of weight ≥ 2 lies in $\mathcal{Z}_2(V)$. For this consider the *Alexander-Whitney map*

$$\text{AW} : \bar{B}_*(V) \to \bar{B}_*(V) \otimes \bar{B}_*(V)$$

$$\text{AW}[v_1 \mid \cdots \mid v_k] = \sum_{j=0}^{k} [v_1 \mid \cdots \mid v_j] \otimes [v_{j+1} \mid \cdots \mid v_k]$$

and notice that the composite

$$m : \bar{B}_*(V) \xrightarrow{\text{AW}} \bar{B}_*(V) \otimes \bar{B}_*(V) \xrightarrow{EZ} \bar{B}_*(V \times V) \xrightarrow{+} \bar{B}_*(V)$$

is chain homotopic to the map μ_2. Since this chain homotopy is natural in V it preserves the "rank"-filtration and hence m and μ_2 induces the same map on $\mathcal{P}_T(V)$. We conclude that

$$(4.13) \qquad \mu_2[v_1 \mid \cdots \mid v_k] = \sum_{j=0}^{k} [v_1 \mid \cdots \mid v_j] \wedge [v_{j+1} \mid \cdots \mid v_k]$$

and hence

$$(\mu_2 - 2)[v_1 \mid \cdots \mid v_k] \in \mathcal{Z}_2(V).$$

It follows that if $x \in \mathcal{P}_T(V)$ has weight $q > 1$ then

$$\mu_2(x) - 2x = (2^q - 2)x \in \mathcal{Z}_2(V)$$

so that $x \in \mathcal{Z}_2(V)$. $\qquad\qquad\qquad\qquad\qquad\qquad\qquad\qquad\square$

The main result on the structure of $\mathcal{P}_T(V)$ is the following reformulation of a theorem of [**Jessen-Thorup, 1978**] and [**Sah, 1979**] which we shall state without proof (cf. [**Dupont, 1982**] and [**Morelli, 1993**]):

THEOREM 4.14. *For V a vector space of dimension n we have*

a) $\tilde{H}_p(\mathcal{T}(V), \wedge_F^q(\mathfrak{g})) = 0$ *for* $p + q \neq n - 1, p \geq -1, q \geq 0$.

b) *The natural map* $\rho : \wedge_{\mathbb{Q}}^q(\mathfrak{g}) \to \wedge_F^q(\mathfrak{g})$ *induces isomorphisms*

$$\rho_* : \tilde{H}_{n-q-1}(\mathcal{T}(V), \wedge_{\mathbb{Q}}^q(\mathfrak{g})) \xrightarrow{\cong} \tilde{H}_{n-q-1}(\mathcal{T}(V), \wedge_F^q(\mathfrak{g}))$$

for $q = 1, 2, \ldots$.

Analogous to theorem 3.5 we now have:

COROLLARY 4.15. *There is a natural isomorphism*

$$\mathcal{P}_T(V) \cong \bigoplus_{q=1}^{n} \mathcal{D}^q(V)$$

where $\mathcal{D}^q(V) = \hat{H}_p(\mathcal{T}(V), \wedge_F^q(\mathfrak{g})), p = n - q - 1 \geq 0$ *is the* F-*subspace*

$$\mathcal{D}^q(V) =$$

$$\left\{ x_\Phi \in \bigoplus_{\Phi = (U_0 \supset \cdots \supset U_p)} \wedge_F^q(U_p)_\Phi \, \Big| \, \sum_{U_i} x_{U_0 \supset \cdots \supset \hat{U}_i \supset U_p} = 0, \quad i = 0, \ldots, p \right\}.$$

and $\mathcal{D}^n = \wedge_F^n(V)$.

Remark: Notice that $\wedge_F^q(U_p)$ is 1-dimensional so if we choose a volume function $v_\Phi \in \wedge_F^q(U_p^*)$ for each strict flag Φ ($v_\phi \in \wedge_F^n(V^*)$ for the "empty" flag) then for $x \in \mathcal{P}_T(V)$ the component $x_\Phi \in \wedge^q(U_p)_\Phi$ is determined by the corresponding *Hadwiger invariant* H_Φ defined as the composite

$$\mathcal{P}_T(V) \cong \bigoplus_q \mathcal{D}^q(V) \xrightarrow{\text{proj}_\Phi} \wedge^q(U_p)_\Phi \xrightarrow{v_\Phi} F$$

Theorem 4.14 thus contains the statement that the Hadwiger invariants determine the translational scissors congruence class. But it also expresses the relations between them. This will be useful in the next chapter. *Geometrically* the Hadwiger invariant H_Φ corresponding to the strict flag $\Phi = (U_0 \supset \cdots \supset U_p)$ and volume function v_Φ is defined on a simplex $\sigma = (a_0, \ldots, a_n) \in V \backslash C_n(V)$ as follows: σ is called *incident* with Φ, written $\sigma \| \Phi$ if for some permutation π, $U_i^\pi = \text{span}\{a_{\pi(i+1)}, \ldots, a_{\pi(n)}\}$ are parallel to $U_i, i = 0, \ldots, p$. Then

$$(4.16) \qquad H_\Phi(\sigma) = \begin{cases} \sum_\pi \text{sign } \pi \; v_\Phi(a_{\pi(i+1)}, \ldots, a_{\pi(n)}) & \text{for } \sigma \parallel \Phi \\ 0 & \text{for } \sigma \nparallel \Phi \end{cases}$$

where $v_\Phi(a_{\pi(i+1)}, \ldots, a_{\pi(n)})$ denotes the (signed) volume in U_p^π corresponding to v_Φ.

EXAMPLE 4.17. For $x = [\Delta]$ a simplex $\Delta = (a_0, a_1, a_2, a_3)$ in E^3 and $q = 1$, the components x_Φ for each pair $\Phi = (U_0 \supset U_1)$ of a plane and a line incident with Δ can be read off the following figure:

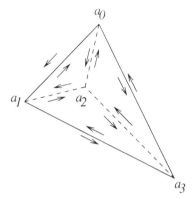

Finally let us state the following Lusztig exact sequences. The proof is similar to the one for the Steinberg modules in chapter 3 (for details see [**Dupont, 1982**]):

THEOREM 4.18. *Let V be a vector space of dimension n over F. Then for $q = 1, 2, \ldots, n$ there is an exact sequence*

$$0 \to \mathcal{D}^q(V) \xrightarrow{\partial} \bigoplus_{U^{n-1}} \mathcal{D}^q(U^{n-1}) \xrightarrow{\partial} \cdots \xrightarrow{\partial} \bigoplus_{U^{q+1}} \mathcal{D}^q(U^{q+1}) \to$$

$$\to \bigoplus_{U^q} \wedge_F^q(U^q) \to \wedge_F^q(U) \to 0$$

where $\partial : \mathcal{D}^q(U^i) \to \mathcal{D}^q(U^{i-1})$ is induced by sending a strict flag $(U_0 \supset \cdots \supset U_{i-q-1})$ in U^i to $(U_1 \supset \cdots \supset U_{i-q-1})$ in $U^{i-1} = U_0$.

Remark 1: Similarly to the "suspension sequence" theorem 3.13 we have another exact sequence (see [**Dupont, 1982**])

(4.19)
$$0 \to \mathcal{D}^q(V) \to \bigoplus_{U^q} \wedge_F^q(U^q) \otimes \mathrm{St}(V/U^q) \to \bigoplus_{U^{q+1}} \wedge_F^q(U^{q+1}) \otimes \mathrm{St}(V/U^{q+1}) \to$$

$$\to \cdots \to \bigoplus_{U^{n-1}} \wedge_F^q(U^{n-1}) \to \wedge_F^q(V) \to 0$$

where for W an m-dimensional vector space $\mathrm{St}(W) = \tilde{H}_{m-2}(\mathcal{T}(W), \mathbb{Z})$.

Remark 2: Notice that all these sequences are sequences of $\mathrm{Gl}(V)$-modules. We shall use this in the next chapter.

CHAPTER 5

Euclidean scissors congruences

Now let us return to the case $X = E^n$ and G a subgroup of $E(n)$, the group of Euclidean isometries, and we want to study the scissors congruence group $\mathcal{P}(E^n, G)$. In particular in dimension 3 we shall establish the exact sequence a) in theorem 1.7. Let us assume that G contains $T(n)$, the group of translations and let $\bar{G} = G/T(n) \subseteq O(n)$. Hence by proposition 2.1

$$(5.1) \qquad \mathcal{P}(E^n, G) \cong H_0(\bar{G}, \mathcal{P}_T(\mathbb{R}^n)^t)$$

and by corollary 4.15 we have a natural isomorphism

$$\mathcal{P}_T(\mathbb{R}^n)^t \cong \bigoplus_{q=1}^{n} \mathcal{D}^q(\mathbb{R}^n)^t.$$

Notice that we have twisted the action on the right hand side of (5.1) in view of corollary 2.11. Hence by (5.1)

$$(5.2) \qquad \mathcal{P}(E^n, G) \cong \bigoplus_{q=1}^{n} H_0(\bar{G}, \mathcal{D}^q(\mathbb{R}^n)^t)$$

COROLLARY 5.3. *Suppose $\bar{G}$ contains* $-$ id. *Then*

$$\mathcal{P}(E^n, G) \cong \bigoplus_{q \equiv n \pmod 2} H_0(\bar{G}, \mathcal{D}^q(\mathbb{R}^n)^t)$$

Proof: Clearly $-$ id acts as the identity on the Tits-complex $\mathcal{T}(\mathbb{R}^n)$ and hence as $(-1)^{q+n}$ on $\mathcal{D}^q(\mathbb{R}^n)^t$. Hence, if $q \neq n$ mod 2, the group $H_0(\bar{G}, \mathcal{D}^q(\mathbb{R}^n)^t)$ is annihilated by 2 and is therefore 0 since it is a rational vector space. $\square$

Remark: In particular corollary 5.3 applies for the group generated by all "point reflections" (or "symmetries"), i.e. by isometries of the form

$S_p, p \in E^n$, given by

$$S_p(x) = p - (x - p) = 2p - x, \quad x \in \mathbb{R}^n.$$

For example in the case of $n = 2$ we then have

$$\mathcal{P}(E^n, G) \cong H_0(\bar{G}, \mathcal{D}^2(\mathbb{R}^2)) = \wedge^2_{\mathbb{R}}(\mathbb{R}^2) \cong \mathbb{R}$$

and the isomorphism is induced by the area function.

Before we proceed we need two simple but useful lemmas about group homology. The first generalizes the use of -id in the proof above and is referred to as the "Center kills lemma":

LEMMA 5.4. *Let M be a (left) $R[G]$-module (R any commutative ring) and let $\gamma \in G$ be a central element such that for some $r \in R, \gamma x = rx, \forall x \in M$. Then $(r - 1)$ annihilates $H_*(G, M)$.*

Proof: For $\gamma \in G$ any element there is an induced endomorphism γ_* of $H_*(G, M)$ given by inner conjugation by γ in G and the action on M. Explicitly in terms of the bar complex $\bar{B}_* G \otimes_R M$,

$$\gamma_*([g_1| \ldots |g_q] \otimes x) = [\gamma g_1 \gamma^{-1}| \ldots |\gamma g_q \gamma^{-1}] \otimes \gamma x$$

and the chain homotopy to the identity is given by

$$s([g_1| \ldots |g_q] \otimes x) = \sum_{i=0}^{q} (-1)^i [g_1| \ldots |g_i|\gamma^{-1}|\gamma g_{i+1} \gamma^{-1}| \ldots |\gamma g_q \gamma^{-1}] \otimes x.$$

Hence if $\gamma \in G$ is a central element acting on M by multiplication by $r \in R$ then

$$\gamma_*([g_1 | \cdots | g_q] \otimes x) = [g_1 | \cdots | g_q] \otimes rx$$

induces both the identity and multiplication by r in homology. This proves the lemma. $\qquad\square$

The second lemma relates the homology of a group with that of a subgroup and is usually called "Shapiro's lemma":

LEMMA 5.5. *Let $H \subseteq G$ be a subgroup and M a (left) $R[G]$-module (R a commutative ring). Then there is a natural isomorphism*

$$H_*(H, M) \cong H_*(G, R[G] \bigotimes_{R[H]} M)$$

Proof: The inclusion $H \subset G$ induces a commutative diagram of chain complexes with vertical isomorphisms:

$$
\begin{array}{ccccc}
B_*(H) \otimes_{R[H]} M & \longrightarrow & B_*(G) \otimes_{R[G]} R[G] \otimes_{R[H]} M & \overset{\cong}{\longrightarrow} & B_*(G) \otimes_{R[H]} M \\
\cong \big\uparrow & & \cong \big\uparrow & & \\
\bar{B}_*(H) \otimes_R M & \longrightarrow & \bar{B}_*(G) \otimes_R \left(R[G] \otimes_{R[H]} M \right) & &
\end{array}
$$

[Here $B_*(H)$ and $B_*(G)$ are considered as *right* modules via the inversion map.] Since $B_*(G)$ is a chain complex of free $R[H]$-modules the horizontal maps above induce a homology isomorphism by the comparison theorem (See e.g. [**MacLane, 1963**, chapter III §6].)

Remark: An explicit inverse is given on the chain level as follows: Choose representatives for the cosets G/H and for $g \in G$ let $\hat{g} \in G$ denote the representatives of gH, that is, $\hat{g}_1 = \hat{g}_2 \Leftrightarrow g_1^{-1} g_2 \in H$. Now define $\rho : \bar{B}_*(G) \bigotimes (R[G] \bigotimes_{R[H]} M) \to \bar{B}(H) \bigotimes_R M$ by

$$
\rho([g_1 \mid g_2 \mid \cdots \mid g_q] \otimes g \otimes x) = [\hat{g}^{-1} g_1 \hat{z}_1 \mid \hat{z}_1^{-1} g_2 \hat{z}_2 \mid \cdots \mid \hat{z}_{q-1}^{-1} g_q \hat{z}_q] \otimes (\hat{g}^{-1} g) x
$$

where $z_j = g_j^{-1} \ldots g_1^{-1} g, j = 1, \ldots, q$.

Now let us return to the calculation of the scissors congruence group $\mathcal{P}(E^n) = \mathcal{P}(E^n, E(n))$. By corollary 5.3

$$
(5.6) \qquad \mathcal{P}(E^n) \cong \bigoplus_{q \equiv n \ (\mathrm{mod}\ 2)} H_0(O(n), \mathcal{D}^q(\mathbb{R}^n)^t)
$$

Here $H_0(O(n), \mathcal{D}^n(\mathbb{R}^n)^t) = H_0(O(n), \wedge_{\mathbb{R}}^n(\mathbb{R}^n)) \cong \mathbb{R}$ and the isomorphism is given by the volume function. For the calculation of the other summands in (5.6) the Lusztig exact sequences in chapter 4 relate the homology of $O(n)$ for the module $\mathcal{D}^q(\mathbb{R}^n)^t$ to the homology for the module $\wedge_{\mathbb{R}}^q(\mathbb{R}^n)^t$ via the homology groups for smaller orthogonal groups. Thus for the other terms in the sequence in theorem 4.18 we have

LEMMA 5.7. *For* $n \geq \ell \geq q$

a)

$$
H_* \left(O(n), \left[\bigoplus_{U^\ell} \mathcal{D}^q(U^\ell) \right]^t \right) \cong H_*(O(\ell), \mathcal{D}^q(\mathbb{R}^l)^t) \bigotimes_{\mathbb{Z}} H_*(O(n - \ell), \mathbb{Z}^t).
$$

b) *In particular this vanishes if $\ell \not\equiv n$* (mod 2).

Proof:

a) As $O(n)$-modules

$$\left[\bigoplus_{U^\ell} \mathcal{D}^q(U^\ell)\right]^t \cong \mathbb{Z}[O(n)] \bigotimes_{\mathbb{Z}[O(\ell)\times O(n-l)]} \mathcal{D}^q(\mathbb{R}^\ell)^t \otimes \mathbb{Z}^t$$

Hence a) follows from Shapiro's lemma (5.5) and the Künneth theorem (4.5).

b) clearly follows from a) and the "center kills" lemma (5.4) applied to the element $\gamma \in O(\ell) \times O(n - \ell)$ given by

$$\gamma(x_1, \ldots, x_n) = (x_1, \ldots, x_\ell, -x_{\ell+1}, \ldots, -x_n)$$

□

Now let us apply this in low dimensions:

EXAMPLE 5.8. $n = 1, 2$. By (5.6) we obtain

$$\mathcal{P}(E^1) \cong H_0(O(1), \wedge_\mathbb{R}^1(\mathbb{R}^1)^t) \cong \mathbb{R},$$
$$\mathcal{P}(E^2) \cong H_0(O(2), \wedge_\mathbb{R}^2(\mathbb{R}^2)^t) \cong \mathbb{R},$$

and the isomorphisms are given by the "length" and "area" functions respectively.

EXAMPLE 5.9. $n = 3$:

$$\mathcal{P}(E^3) \cong H_0(O(3), \mathcal{D}^1(\mathbb{R}^3)^t) \oplus H_0(O(3), \wedge_\mathbb{R}^3(\mathbb{R}^3)^t)$$
$$\cong H_0(O(3), \mathcal{D}^1(\mathbb{R}^3)^t) \oplus \mathbb{R}$$

where the projection onto the last summand is given by the volume function. Note also that by proposition 4.12

$$\mathcal{D}^1(\mathbb{R}^3)^t \cong \mathcal{P}_T(\mathbb{R}^3)/\mathcal{Z}_2(\mathbb{R}^3)$$

where $\mathcal{Z}_2(\mathbb{R}^3)$ is generated by products of 1- and 2- simplices, i.e. by prisms. Hence the sequence a) in theorem 1.7 follows from the following:

THEOREM 5.10. *There is an exact sequence*

$$0 \to H_2(O(3), (\mathbb{R}^3)^t) \to H_0(O(3), \mathcal{D}^1(\mathbb{R}^3)^t) \to \mathbb{R} \otimes_{\mathbb{Z}} \mathbb{R}/\mathbb{Z} \to$$
$$\to H_1(O(3), (\mathbb{R}^3)^t) \to 0$$

Furthermore the composite map

$$D : \mathcal{P}(E^3) \to H_0(O(3), \mathcal{D}^1(\mathbb{R}^3)^t) \to \mathbb{R} \otimes_{\mathbb{Z}} \mathbb{R}/\mathbb{Z}$$

is the Dehn-invariant.

Proof: We split the Lusztig sequence (4.18) into two exact sequences

(5.11.a) $$0 \to \mathcal{D}^1(\mathbb{R}^3) \to \bigoplus_{U^2} \mathcal{D}^1(U^2) \to K \to 0$$

(5.11.b) $$0 \to K \to \bigoplus_{U^1} U^1 \to \mathbb{R}^3 \to 0.$$

Then by (5.11.a) and lemma 5.7 b)

$$H_0(O(3), \mathcal{D}^1(\mathbb{R}^3)^t) \cong H_1(O(3), K^t)$$

and

$$H_0(O(3), K^t) = 0.$$

Also

$$H_*\left(O(3), \left[\bigoplus_{U^1} U^1\right]^t\right) \cong H_*(O(1), (\mathbb{R}^1)^t) \bigotimes_{\mathbb{Z}} H_*(O(2), \mathbb{Z}^t)$$
$$\cong \mathbb{R} \bigotimes_{\mathbb{Z}} H_*(O(2), \mathbb{Z}^t).$$

By proposition 4.7, iii)

$$H_*(SO(2), \mathbb{Z})/\text{Torsion} \cong \wedge^*_{\mathbb{Q}}(SO(2)/\text{Torsion}) \cong \wedge^*_{\mathbb{Q}}(\mathbb{R}/\mathbb{Q})$$

where we have identified a rotation of angle θ with $\theta/2\pi \in \mathbb{R}/\mathbb{Z}$. Here the action of $O(2)/SO(2) = \mathbb{Z}/2$ is induced by $\theta \mapsto -\theta$. In particular

$$H_i\left(O(3), \left[\bigoplus_{U^1} U^1\right]^t\right) = \begin{cases} \mathbb{R} \otimes_{\mathbb{Z}} (\mathbb{R}/\mathbb{Z}) & i = 1 \\ 0 & i = 0, 2. \end{cases}$$

We now obtain the desired exact sequence from the exact sequence (5.11.b). We leave the identification of the Dehn-invariant as an exercise [note that there is a factor $\frac{1}{2}$ which however is immaterial for exactness.]

$\square$

Thus we have established the exact sequence a) in theorem 1.7. Hence we have proved the following reformulation of Sydler's theorem (1.4):

COROLLARY 5.12. $D : \mathcal{P}(E^3)/\mathcal{Z}(E^3) \to \mathbb{R} \bigotimes_{\mathbb{Z}} \mathbb{R}/\mathbb{Z}$ is injective if and only if $H_2(O(3),(\mathbb{R}^3)^t) = 0$.

EXAMPLE 5.13. $n = 4$:

$$\mathcal{P}(E^4) \cong H_0(O(4),\mathcal{D}^2(\mathbb{R}^4)^t) \oplus H_0(O(4),\wedge^4_{\mathbb{R}}(\mathbb{R}^4)^t)$$
$$\cong H_0(O(4),\mathcal{D}^2(\mathbb{R}^4)^t) \oplus \mathbb{R}$$

where again the projection onto the last summand is given by the volume function. For the calculation of the first summand we again consider the appropriate Lusztig sequence (4.18) and similarly to theorem 5.10 we obtain an exact sequence

(5.14)

$$0 \to H_2(O(4),\wedge^2(\mathbb{R}^4)^t) \to H_0(O(4),\mathcal{D}^2(\mathbb{R}^4)^t) \to \wedge^2_{\mathbb{R}}(\mathbb{R}^2) \bigotimes_{\mathbb{Z}} \mathbb{R}/\mathbb{Z} \to$$

$$\to H_1(O(4),(\mathbb{R}^4)^t) \to 0$$

Also one checks that (except for a factor $\frac{1}{2}$) the composite map

$$D : \mathcal{P}(E^4) \to H_0(O(4),\mathcal{D}^2(\mathbb{R}^4)^t) \to \wedge^2_{\mathbb{R}}(\mathbb{R}/\mathbb{Z}) \cong \mathbb{R} \otimes \mathbb{R}/\mathbb{Z}$$

is the Dehn invariant, i.e. for P a 4-dimensional polytope,

(5.15) $$D(P) = \sum_A \text{Area } (A) \otimes \theta(A)/\pi$$

where A runs through all 2-dimensional faces of P with area Area(A) and dihedral angle $\theta(A)$. For the computation of the homology groups in (5.14) we note that by the Künneth theorem,

$$H_i(SO(4),\wedge^2_{\mathbb{R}}(\mathbb{R}^4)) \cong H_i(\text{Spin}(4),\wedge_+ \oplus \wedge_-)$$
$$\cong H_i(S_+,\wedge_+) \oplus H_i(S_-,\wedge_-), \text{ for } i \leq 2,$$

where $\text{Spin}(4) \cong S_+ \times S_-$ with $S_\pm \simeq \text{Spin}(3)$ and $\wedge^2_{\mathbb{R}}(\mathbb{R}^3) \cong \wedge_+ \oplus \wedge_-$. Here $\wedge_\pm \cong \mathbb{R}^3$ and $S_\pm$ acts via the usual action of $O(3)$ on $\wedge_\pm$ but trivially on $\wedge_{\mp}$ so that in particular $H_1(S_{\pm}, \wedge_{\mp}) = 0$ since $\mathrm{Spin}(3)$ is a perfect group. Furthermore change of orientation interchanges S_+ and S_- (respectively $\wedge_+$ and $\wedge_-$). It follows that for $i \leq 2$

$$H_i(O(4), \wedge_{\mathbb{R}}^2(\mathbb{R}^4)^t) \cong H_i(SO(3), \mathbb{R}^3) \cong H_i(O(3), (\mathbb{R}^3)^t)$$

(since $O(3) \cong SO(3) \times \{\pm 1\}$). If now we let $\mathcal{Z}(E^4) \subseteq \mathcal{P}(E^4)$ be the subgroup generated by products of 1- and 2-simplices it is easy to see that $\mathrm{Vol} : \mathcal{Z}(E^4) \to \mathbb{R}$ is an isomorphism, and we conclude the following:

COROLLARY 5.16. $D : \mathcal{P}(E^4)/\mathcal{Z}(E^4) \to \mathbb{R} \bigotimes_{\mathbb{Z}} \mathbb{R}/\mathbb{Z}$ is injective if and only if $H_2(O(3), (\mathbb{R}^3)^t) = 0$.

Thus we have proved the result of [**Jessen, 1972**] that Sydler's theorem is true in E^4 if and only if it holds in E^3.

EXAMPLE 5.17. $n = 5$: As before

$$\mathcal{P}(E^5) \cong H_0(O(5), \mathcal{D}^1(\mathbb{R}^5)^t) \oplus H_0(O(5), \mathcal{D}^3(\mathbb{R}^5)^t) \oplus \mathbb{R}.$$

Again the projection onto the last summand is just the volume function and one can show that the middle summand is detected by a Dehn-invariant similarly as above. There is also a generalized Dehn invariant associated to the first factor but the kernel of this involves the computation of $H_4(SO(5), \mathbb{R}^5)$ which seems rather difficult.

Remark: It follows from the above calculations that (granted Sydler's theorem) there are isomorphisms

$$\mathcal{P}(E^1) \cong \mathcal{P}(E^2), \quad \mathcal{P}(E^3) \cong \mathcal{P}(E^4)$$

and in both cases the isomorphism is given by multiplication by the 1-simplex Δ^1 of unit length. It is a question by B. Jessen if in general this gives an isomorphism $\mathcal{P}(E^{2m+1}) \to \mathcal{P}(E^{2m+2})$. More specifically he asked if a 6-dimensional polytope is scissors congruent to a product of a 1-simplex and a 5-dimensional polytope (cf. [**Sah, 1979**, chapter 7]). Notice that in terms of the decomposition (5.6) the multiplication map with Δ^1 induces a well-defined map $\mathcal{D}^q(\mathbb{R}^n)^t \to \mathcal{D}^{q+1}(\mathbb{R}^{n+1})^t$ for all n and $q = 1, \dots, n$. A more general question is if this induces an isomorphism on the level of $H_0(O(n), -)$. This is known to be true for $q = n$ and $q = n - 2$ for all n (see [**Dupont, 1982**]).

CHAPTER 6

Sydler's theorem and non-commutative differential forms

We will now sketch a direct homological proof of the vanishing result

$$H_2(\mathrm{SO}(3), \mathbb{R}^3) = 0$$

and thereby prove Sydler's theorem (1.4) and the corresponding result in E^4 (cf. corollary 5.16). We refer to [**Dupont-Sah, 1990**] for the details. Another proof has been given by [**Cathelineau, 1998**].

There are two main ingredients in our proof: One is the following general vanishing theorem for which we defer the proof to chapter 9 (theorem 9.11).

THEOREM 6.1 $H_i(\mathrm{O}(n), \wedge_{\mathbb{Q}}^j(\mathbb{R}^n)) = 0$ if $i + j \leq n, \quad j > 0.$

Note that this does not directly give the desired vanishing for $n = 3$ and $j = 1$ since the coefficients are untwisted (so that in this case the homology vanishes trivially by "center kills"). Indeed we shall use the case $n = 4$ and $j = 2$ for our proof.

The other ingredient is the calculation of the Hochschild homology of the real quaternions $\mathbb{H}$ considered as an algebra over $\mathbb{Q}$ together with the formulation of this homology in terms of *non-commutative differential forms* (due to M. Karoubi [**Karoubi, 1986**]). But first let us recall for A any *commutative* algebra over $\mathbb{Q}$ with a unit element the notion of *Kähler differentials* $\Omega_A^n, n = 0, 1, 2, \ldots$: Let $I(A) \subseteq A \otimes A \, (= A \bigotimes_{\mathbb{Q}} A)$ be the kernel of the multiplication map

$$A \otimes A \to A$$

and put $\Omega_A^1 = I(A)/I(A)^2$, where $I(A)^2$ is the two sided ideal in $A \otimes A$ generated by the image $I(A) \otimes I(A) \to I(A)$. Then $\Omega_A^0 = A$ and $\Omega_A^n = \wedge_A^n(\Omega_A^1), n = 1, 2, \cdots$. Furthermore for $a \in A$ we let $da \in \Omega_A^1$ denote the image of $1 \otimes a + a \otimes 1 \in A \otimes A$. In this notation we shall prove:

>   THEOREM 6.2.   a) $H_1(\mathrm{SO}(3), \mathbb{R}^3) \cong \Omega_{\mathbb{R}}^1$ and in the exact se-
>   quence of theorem 1.7, a) the last map corresponds to the map
> $J : \mathbb{R} \otimes \mathbb{R}/\mathbb{Z} \to \Omega_{\mathbb{R}}^1$ given by
>
> $$J(\ell \otimes \theta/\pi) = \ell \frac{d\cos\theta}{\sin\theta}$$

b) $H_2(\mathrm{SO}(3), \mathbb{R}^3) = 0$.

Remark: The statement in a) determines the image of the Dehn-invariant as the kernel of J and is a reformulation of a classical result of [**Jessen, 1968**].

In the following A is a general associative (but not necessarily commutative) algebra over $\mathbb{Q}$ and all tensor products are over $\mathbb{Q}$ unless otherwise specified. Recall that the *Hochschild homology* $HH_*(A)$ of A is the homology of the chain complex (A^*, b) which in degree n is $A^{\otimes(n+1)} = A \otimes A \otimes \cdots \otimes A$ $(n + 1$ factors) with boundary map b given by

(6.3)
$$b(a_0 \otimes \cdots \otimes a_n) = \sum_{i=0}^{n-1}(-1)^i a_0 \otimes \cdots \otimes a_i a_{i+1} \otimes \cdots \otimes a_n +$$
$$+ (-1)^n a_n a_0 \otimes \cdots \otimes a_{n-1}.$$

That is

$$HH_n(A) = H_n(A^*, b).$$

However A^* is a *simplicial abelian group* with face operators

$$\varepsilon_i : A^{\otimes(n+1)} \to A^{\otimes n}$$

given by

$$\varepsilon_i(a_0 \otimes \cdots \otimes a_n) = \begin{cases} a_0 \otimes \cdots \otimes a_i a_{i+1} \otimes \cdots \otimes a_n, & i = 0, \ldots, n-1, \\ a_n a_0 \otimes \cdots \otimes a_{n-1}, & i = n. \end{cases}$$

and we consider the set of *non-commutative differential forms*

$$\Omega_n(A) = \bigcap_{i=0}^{n-1} \ker(\varepsilon_i), \quad \Omega_0(A) = A.$$

Then $b \mid \Omega_n(A) = (-1)^n \varepsilon_n$ and it follows (see e.g. [**May, 1967**, theorem 22.1]) that

(6.4) $$HH_n(A) \cong H_n(\Omega_*(A), b).$$

Notice that $\Omega_*(A)$ is a graded algebra with the multiplication

$$(a_0 \otimes \cdots \otimes a_n)(b_0 \otimes \cdots \otimes b_m) = a_0 \otimes \cdots \otimes a_n b_0 \otimes \cdots \otimes b_m.$$

Again for $a \in A$ we put $da = 1 \otimes a - a \otimes 1 \in \Omega_1(A)$ and it is easy to see that

$$\Omega_n(A) \cong A \otimes (A/\mathbb{Q}) \otimes \cdots \otimes (A/\mathbb{Q})$$

via the map

$$\omega = a_0 da_1 \ldots da_n \quad \longleftarrow\!\!\!\mid a_0 \otimes a_1 \otimes \cdots \otimes a_n.$$

We can then extend $d : \Omega_n(A) \to \Omega_{n+1}(A)$ for all n by

$$d(a_0 da_1 \ldots da_n) = da_0 da_1 \ldots da_n$$

such that in general

$$d(\omega \cdot \omega') = (d\omega)\omega' + (-1)^{\deg \omega} \omega \cdot d\omega'.$$

That is $(\Omega_*(A), d)$ is a *differential graded algebra*. In this notation $b : \Omega_n(A) \to \Omega_{n-1}(A)$ is given by

$$b(\omega da) = (-1)^n (a\omega - \omega a)$$

and we put

$$I_n(A) = \ker(b : \Omega_n(A) \to \Omega_{n-1}(A))$$
$$B_n(A) = \operatorname{im}(b : \Omega_{n+1}(A) \to \Omega_n(A))$$

so that

$$HH_n(A) \cong I_n(A)/B_n(A).$$

If A is *commutative* then $HH_1(A) = I_1(A)/B_1(A) = \Omega_A^1$ is the set of Kähler differentials defined above and the usual *shuffle product* defines a map

$$\gamma : \Omega_A^n \to HH_n(A).$$

For $A = K$ a field we quote without proof the following well-known theorem of [**Hochschild-Kostant-Rosenberg, 1962**].

THEOREM 6.5. *If K is a field of characteristic 0 then*

$$HH_n(K) \cong \Omega_K^n, \quad \text{for all } n,$$

via the map γ.

COROLLARY 6.6. *The inclusion of $\mathbb{Q}$-algebras $\mathbb{R} \subseteq \mathbb{H}$ induces an isomorphism*

$$\Omega_{\mathbb{R}}^n \cong HH_n(\mathbb{R}) \cong HH_n(\mathbb{H}) \text{ for all } n.$$

Proof: By the Künneth theorem

$$HH_*(\mathbb{H}) \cong HH_*(\mathbb{H}_0) \otimes HH_*(\mathbb{R})$$

where $\mathbb{H}_0 = \mathrm{span}_{\mathbb{Q}}\{1, i, j, k\}$ is the quaternion algebra over $\mathbb{Q}$ so that $\mathbb{H} \cong \mathbb{R} \otimes \mathbb{H}_0$. Hence is suffices to show that $HH_0(\mathbb{H}_0) \cong \mathbb{Q}$ and $HH_n(\mathbb{H}_0) = 0$ for $n > 0$. However $\mathbb{H}_0 \otimes \mathbb{H}_0 \cong M_4(\mathbb{Q})$ the full 4×4 matrix algebra over $\mathbb{Q}$ and since Hochschild homology is a Morita invariant (see e.g. [**Kassel, 1982**]) we conclude again from the Künneth theorem that

$$HH_*(\mathbb{H}_0) \otimes HH_*(\mathbb{H}_0) \cong HH_*(M_4(\mathbb{Q})) \cong HH_*(\mathbb{Q}) = \begin{cases} \mathbb{Q} & * = 0 \\ 0 & * > 0 \end{cases}$$

which implies the desired conclusion. $\qquad\qquad\qquad\qquad\qquad\qquad\square$

Next let us return to the calculation of $H_*(\mathrm{SO}(3), \mathbb{R}^3)$ but reformulated in terms of quaternions: As usual

$$\mathrm{Spin}(3) = \mathrm{Sp}(1) = \{q \in \mathbb{H} \mid q\bar{q} = 1\}$$

and the double covering $\rho : \mathrm{Spin}(3) \to \mathrm{SO}(3)$ is given

$$\rho(q)(v) = qv\bar{q}, \quad q \in \mathrm{Sp}(1) \quad v \in \mathrm{span}\{i, j, k\} \subseteq \mathbb{H}$$

Similarly (as mentioned in chapter 5 with different notation) $\mathrm{Spin}(4) \cong S_1 \times S_2$ with $S_i \cong \mathrm{Sp}(1), i = 1, 2$, and $\sigma : \mathrm{Spin}(4) \to \mathrm{SO}(4)$ is given by

$$\sigma(q_1, q_2)(v) = q_1 v \bar{q}_2 \quad q_1, q_2 \in \mathrm{Sp}(1) \quad v \in \mathbb{H}.$$

We extend this to $\sigma : \mathrm{Pin}(4) \to \mathrm{O}(4)$, where $\mathbb{Z}/2$ interchanges the factors and acts on $\mathbb{H}$ by the quaternionic conjugation $v \mapsto \bar{v}$. Now theorem

6.1 for $n = 4$ can be reformulated (using the Hochschild-Serre spectral sequence for the double covering σ):

$$(6.7) \qquad H_\ell(\mathrm{Pin}(4), \wedge_\mathbb{Q}^2(\mathbb{H})) = 0 \text{ for } \ell = 0, 1, 2.$$

Here $\wedge_\mathbb{Q}^2(\mathbb{H}) \cong (\mathbb{H} \otimes \mathbb{H})^-$ the (-1)-eigenspace for the involution $\tau(a_0 \otimes a_1) = \bar{a}_1 \otimes \bar{a}_0$ via the map $a_0 \wedge a_1 \leftrightarrow \frac{1}{2}(a_0 \otimes \bar{a}_1 - a_1 \otimes \bar{a}_0)$. This isomorphism is $\mathrm{Pin}(4)$-equivariant if the action on $\mathbb{H} \otimes \mathbb{H}$ is defined by

$$\tilde{\sigma}(q_1, q_2)(a_0 \otimes a_1) = q_1 a_0 \bar{q}_2 \otimes q_2 a_1 \bar{q}_1, \quad q_1, q_2 \in \mathrm{Sp}(1),$$

and $\mathbb{Z}/2$ acts by $a_0 \otimes a_1 \mapsto \bar{a}_0 \otimes \bar{a}_1$. Next the map

$$\varepsilon_0 \oplus (-\varepsilon_1) : \mathbb{H} \otimes \mathbb{H} \to \mathbb{H} \oplus \mathbb{H}$$

sending $a_0 \otimes a_1$ to $(a_0 a_1, -a_1 a_0)$ is $\mathrm{Pin}(4)$-invariant for the action above on the left and the action

$$\tilde{\sigma}(q_1, q_2)(v_1, v_2) = (\rho(q_1)(v_1), \rho(q_2)(v_2))$$

and $\mathbb{Z}/2$ acts by $(v_1, v_2) \mapsto (-\bar{v}_1, -\bar{v}_1)$. Also we let τ act on the right by $\tau(v_1, v_2) = (\bar{v}_2, \bar{v}_2)$ and denote the (± 1)-eigenspaces for this involution by upper indices $\pm$. Since $HH_0(\mathbb{H}) = HH_0(\mathbb{R}) = \mathbb{R} = \mathbb{H}^+$ we get an exact sequence of $\mathrm{Pin}(4)$-modules

$$(6.8) \qquad 0 \to I_1(\mathbb{H})^- \to (\mathbb{H} \otimes \mathbb{H})^- \xrightarrow{\varepsilon_0 \oplus -\varepsilon_1} \mathbb{H}^- \oplus \mathbb{H}^- \to 0$$

where $\mathbb{H}^- = \mathrm{span}\{i, j, k\}$. Here we have for $0 \le \ell \le 2$:

$$H_\ell(\mathrm{Pin}(4), \mathbb{H}^- \oplus \mathbb{H}^-) \cong H_\ell(\mathrm{Spin}(3), \mathbb{H}^-) \cong H_\ell(\mathrm{SO}(3), \mathbb{R}^3).$$

Hence by (6.7)

$$H_\ell(\mathrm{SO}(3), \mathbb{R}^3) \cong H_{\ell-1}(\mathrm{Pin}(4), I_1(\mathbb{H})^-) \quad \ell = 1, 2,$$

and therefore theorem 6.2 is equivalent to

THEOREM 6.9.   a) *The natural map*

$$I_1(\mathbb{H}) \to HH_1(\mathbb{H}) = HH_1(\mathbb{R}) = \Omega_\mathbb{R}^1$$

induces an isomorphism

$$H_0(\mathrm{Spin}(4), I_1(\mathbb{H})^-) \cong \Omega_\mathbb{R}^1$$

and $\mathrm{Pin}(4)/\mathrm{Spin}(4) = \mathbb{Z}/2$ acts trivially on this.
   b) $H_1(\mathrm{Spin}(4), I_1(\mathbb{H})^-) = 0.$

Proof: (Sketch). First we extend the Spin(4)-action on $\Omega_1(\mathbb{H}) \subseteq \mathbb{H} \otimes \mathbb{H}$ to all of $(\Omega_*(\mathbb{H}), b)$ using the isomorphism

$$\Omega_n(\mathbb{H}) \cong \Omega_1(\mathbb{H}) \otimes (\mathbb{H}/\mathbb{Q})^{\otimes(n-1)}.$$

We thus define

$$\tilde{\sigma}(q_1, q_2)(a_0) = \rho(q_2)(a_0) \quad \text{for} \quad a_0 \in \Omega_0(\mathbb{H}),$$

and

$$\tilde{\sigma}(q_1, q_2)(\omega da_2 \ldots da_n) = \tilde{\sigma}(q_1, q_2)(\omega) d(\rho(q_1)(a_2)) \ldots d(\rho(q_1)(a_n)),$$

for $n \geq 1$. Similarly we extend the involution τ by

$$\tau(a_0) = \bar{a}_0 \text{ for } a_0 \in \Omega_0(\mathbb{H}), \quad \tau(a_0 da_1) = -(d\bar{a}_1)\bar{a}_0$$

and for $n \geq 2$

$$\tau(\omega da_2 \ldots da_n) = -(-1)^{\frac{1}{2}(n-2)(n-3)} \tau(\omega) d\bar{a}_n \ldots d\bar{a}_2.$$

With these definitions τ commutes with the Spin(4)-action and the boundary map so that in particular $\Omega_n, I_n(\mathbb{H})$ and $B_n(\mathbb{H})$ split into (± 1)-eigenspaces for τ as Spin(4)-modules. The following lemma is straight forward.

LEMMA 6.10.   i) Spin(4) *acts trivially on* $HH_*(\mathbb{H})(\cong \Omega_{\mathbb{R}}^*)$.
ii) *On* $HH_n(\mathbb{H})$ *the involution* τ *is* $\tau = (-1)^n \mathrm{id}$.
iii) *For* $S_2 \subseteq \mathrm{Spin}(4)$ *acting on* $\Omega_n(\mathbb{H})$ *we have*

$$H_0(S_2, \Omega_n(\mathbb{H})) \cong \begin{cases} \mathbb{R} & \text{for } n = 0 \\ 0 & \text{for } n > 0 \end{cases}$$

With this we can now prove part a) of theorem 6.9: We consider the exact sequences of Spin(4)-modules

(6.11) $0 \to B_1(\mathbb{H})^- \to I_1(\mathbb{H})^- \to HH_1(\mathbb{H})^- \to 0$

(6.12) $0 \to I_2(\mathbb{H})^- \to \Omega_2(\mathbb{H})^- \overset{b}{\to} B_1(\mathbb{H})^- \to 0$

Here (6.11) gives the exact sequence

$$\to H_0(\mathrm{Spin}(4), B_1(\mathbb{H})^-) \to H_0(\mathrm{Spin}(4), I_1(\mathbb{H})^-) \to$$
$$\to H_0(\mathrm{Spin}(4), HH_1(\mathbb{H})^-) \to 0$$

where the last term is isomorphic to $\Omega_{\mathbb{R}}^1$ by lemma 6.10, i). But the first term is 0 by the sequence (6.12) together with lemma 6.10, iii) for

$n = 2$. This proves a) except for the identification of the map J which follows by direct checking of the maps involved. For the proof of part b) we again obtain from (6.11) the exact sequence

$$(6.13) \quad H_1(\mathrm{Spin}(4), B_1^-) \to H_1(\mathrm{Spin}(4), I_1^-) \to H_1(\mathrm{Spin}(4), HH_1^-)$$

where the last term vanishes by lemma 6.10, i) since $\mathrm{Spin}(4)$ is a perfect group. It remains to prove that the first term in (6.13) vanishes. Now (6.12) yields the exact sequence

$$(6.14) \quad H_1(\mathrm{Spin}(4), \Omega_2^-) \overset{b_*}{\to} H_1(\mathrm{Spin}(4), B_1^-) \to H_0(\mathrm{Spin}(4), I_2^-)$$

where the last term is zero since by lemma 6.10, ii) $I_2^- = B_2^-$, hence

$$H_0(\mathrm{Spin}(4), \Omega_3^-) \overset{b_*}{\to} H_0(\mathrm{Spin}(4), I_2^-)$$

is surjective with the first group equal to zero by lemma 6.10, iii). It follows that the first term (and hence all terms) of the sequence (6.13) vanishes if just $b_* = 0$ in the sequence (6.14). This is shown in [**Dupont-Sah, 1990**], and thus we end our outline of the proof of theorem 6.9 and equivalently of theorem 6.2. □

Thus except for the proof of theorem 6.1 we have ended the proof of Sydler's theorem and this also ends our treatment of Euclidean scissors congruences.

CHAPTER 7

Spherical scissors congruences

We next turn to the study of scissors congruence in spherical space, that is, we want to calculate the scissors congruence group $\mathcal{P}(S^n), n = 0, 1, 2, \ldots$. Recall from chapter 2 (corollary 2.11) that

$$\mathcal{P}(S^n) \cong H_0(O(n+1), \mathrm{Pt}(S^n)^t)$$

Let us first calculate this in low dimensions to illustrate the homological approach:

EXAMPLE 7.1. $n = 0$: $\mathrm{Pt}(S^0)^t$ is clearly a free O(1)-module with one generator (1) so that

$$\mathcal{P}(S^0) = \mathbb{Z} \text{ and } H_i(O(1), \mathrm{Pt}(S^0)^t) = 0 \text{ for } i > 0.$$

Note that theorem 3.12 in this case reduces to the sequence

$$0 \to \mathbb{Z} \to \mathrm{Pt}(S^0)^t \to \mathbb{Z}^t \to 0$$

which gives the exact sequence at H_0-level:

$$0 \to \mathbb{Z} \to \mathcal{P}(S^0) \to \mathbb{Z}/2 \to 0.$$

Here the generator in the first group goes to $(1) - (-1) = 2(1)$ in $\mathcal{P}(S^0) = H_0(O(1), \mathrm{Pt}(S^0)^t)$.

EXAMPLE 7.2. $n = 1$: Theorem 3.12 provides the following short exact sequences

(7.3,i) $$0 \to \mathbb{Z} \to \mathrm{Pt}(S^1)^t \to K_0 \to 0$$

(7.3,ii) $$0 \to K_0 \to \left[\bigoplus_{U^0} \mathrm{Pt}(U^0) \right]^t \to \mathbb{Z}^t \to 0$$

with

$$H_i\left(\mathrm{O}(2),\left[\bigoplus_{U^0}\mathrm{Pt}(U^0)\right]^t\right) \cong H_i(\mathrm{O}(1) \times \mathrm{O}(1), \mathrm{Pt}(S^0)^t \otimes \mathbb{Z}^t)$$

$$\cong H_i(\mathrm{O}(1), \mathbb{Z}^t) \cong \begin{cases} \mathbb{Z}/2, & \text{for } i \text{ even} \\ 0, & \text{for } i \text{ odd}. \end{cases}$$

By (7.3,ii)

$$H_i(\mathrm{O}(2), K_0) \cong H_{i+1}(\mathrm{O}(2), \mathbb{Z}^t)/H_{i+1}(\mathrm{O}(1), \mathbb{Z}^t), \quad i = 0, 1, 2, \ldots,$$

and (7.3,i) gives the exact sequence

$$\cdots \to H_{i+1}(\mathrm{O}(2), K_0) \to H_i(\mathrm{O}(2), \mathbb{Z}) \to H_i(\mathrm{O}(2), \mathrm{Pt}(S^1)^t) \to$$
$$\to H_i(\mathrm{O}(2), K_0) \to H_{i-1}(\mathrm{O}(2), \mathbb{Z}) \to \ldots$$

In particular using the Hochschild-Serre spectral sequence for the extension $1 \to \mathrm{SO}(2) \to \mathrm{O}(2) \to \{\pm 1\} \to 0$ one has

$$H_i(\mathrm{O}(2), K_0) \equiv \begin{cases} H_1(\mathrm{O}(2), \mathbb{Z}^t) \cong \mathbb{R}/\mathbb{Z} & \text{for } i = 0 \\ 0 & \text{for } i = 1. \end{cases}$$

Hence $H_0(\mathrm{O}(2), \mathbb{Z}) = \mathbb{Z}$ injects in $\mathcal{P}(S^1)$ with quotient $\mathbb{R}/\mathbb{Z}$ so indeed $\mathcal{P}(S^1) \cong \mathbb{R}$ given by the "length" function. Notice that $\mathbb{Z} \subseteq \mathcal{P}(S^1)$ is generated by $[S^1]$. Notice also that by the above calculation the map

$$\mathbb{Z}/2 \cong H_1(\mathrm{O}(2), \mathbb{Z}) \to H_1(\mathrm{O}(2), \mathrm{Pt}(S^1)^t)$$

is surjective so the latter group has at most order 2.

For the calculation of $\mathcal{P}(S^n)$ in higher dimensions let us first show the general result of C.-H. Sah [**Sah, 1981**] that $\mathcal{P}(S^{2m}) \cong \mathcal{P}(S^{2m-1})$ for $m = 1, 2, \ldots$. To prove this we consider first the *suspension* map considered in chapter 3 and which clearly induces a map

$$\Sigma : \mathcal{P}(S^{n-1}) \to \mathcal{P}(S^n), \quad n = 1, 2, \ldots.$$

For this we have

THEOREM 7.4.   a) *There is an exact sequence*

$$\to H_1(O(n+1), \mathrm{Pt}(S^n)^t) \overset{h_*}{\to} H_1(O(n+1), \mathrm{St}(S^n)^t) \to \mathcal{P}(S^{n-1}) \overset{\Sigma}{\to} \mathcal{P}(S^n)$$
$$\overset{h_*}{\to} H_0(O(n+1), \mathrm{St}(S^n)^t) \to 0$$

b) *In particular*

$$\Sigma : \mathcal{P}(S^{2m-1}) \to \mathcal{P}(S^{2m}), \quad m = 1, 2, \ldots.$$

is surjective with kernel consisting of elements of order at most 2.

Proof: a) Let us split up the suspension sequence (theorem 3.13) into exact sequences

(7.5,i) $$0 \to K_0 \to \mathrm{Pt}(S^n)^t \overset{h}{\to} \mathrm{St}(S^n)^t \to 0$$

(7.5,ii)
$$\bigoplus_{U^{n-2}} \mathrm{Pt}(U^{n-2})^t \otimes \mathrm{St}(U^{n-2\perp}) \overset{\Sigma}{\to} \bigoplus_{U^{n-1}} \mathrm{Pt}(U^{n-1})^t \otimes \mathrm{St}(U^{n-1\perp}) \to K_0 \to 0$$

where

$$H_* \left(O(n+1), \bigoplus_{U^{n-\ell}} \mathrm{Pt}(U^{n-\ell})^t \otimes \mathrm{St}(U^{n-\ell\perp}) \right)$$
$$\cong H_* \left(O(n-\ell+1) \times O(\ell), \mathrm{Pt}(S^{n-\ell})^t \otimes \mathrm{St}(S^{\ell-1}) \right), \quad \ell = 1, 2.$$

By (7.5,i) we get the exact sequence

$$\overset{h_*}{\to} H_1(O(n+1), \mathrm{St}(S^n)^t) \to H_0(O(n+1), K_0) \to \mathcal{P}(S^n) \to$$
$$\to H_0(O(n+1), \mathrm{St}(S^n)^t) \to 0$$

so it suffices to identify $H_0(O(n+1), K_0)$ with $\mathcal{P}(S^{n-1})$. For this (7.5,ii) yields the exact sequence

(7.6)
$$\mathcal{P}(S^{n-1}) \otimes H_0(O(2), \mathrm{St}(S^1)) \overset{\Sigma}{\to} \mathcal{P}(S^{n-1}) \to H_0(O(n+1), K_0) \to 0$$

But by theorem 3.11 we have the exact sequence of $O(2)$-modules

$$0 \to \mathrm{St}(S^1) \to \bigoplus_{U^0} \mathrm{St}(U^0) \to \mathbb{Z} \to 0$$

with

$$H_*(O(2), \bigoplus_{U^0} \mathrm{St}(U^0)) \cong H_*(O(1) \times O(1), \mathbb{Z})$$

which is a 2-torsion group in all dimensions bigger than 0. Since the inclusion $O(1) \times O(1) \to O(2)$ induces an isomorphism in homology in dimension 0, and is surjective in dimension 1 it follows that

$$H_0(O(2), \mathrm{St}(S^1)) = 0.$$

Hence the first group in the sequence (7.6) is zero which shows the required isomorphism and ends the proof of the first part.

b) By "center kills" for $-\,\mathrm{id} \in O(n+1)$ all $H_0(O(n+1), \mathrm{St}(S^n)^t)$ is annihilated by 2 for n odd. But by Gerling's theorem (corollary 2.5) $\mathcal{P}(S^n)$ is 2-divisible, hence by the sequence in part a) it follows that $H_0(O(n+1), \mathrm{St}(S^n)^t)$ must be zero. $\square$

Notice that $\Sigma = 2c : \mathcal{P}(S^{n-1}) \to \mathcal{P}(S^n)$, where for P an $(n-1)$-dimensional polytope

$$c[P] = [e_{n+1} * P],$$

the *cone* on P with cone point $e_{n+1} = (0, \ldots, 1)$. We then conclude Sah's Cone Theorem:

COROLLARY 7.7. *The cone map* $c : \mathcal{P}(S^{2m-1}) \to \mathcal{P}(S^{2m}), m = 1, 2, \ldots,$ *is an isomorphism.*

Proof: Since $\Sigma(x) = 2c(x) = c(2x)$ for all $x \in \mathcal{P}(S^{2m-1})$ and since Σ is surjective also c is surjective. Now suppose $c(x) = 0$. Then since $\mathcal{P}(S^{2m-1})$ is 2-divisible by Gerling's theorem $x = 2y$ for some $y \in \mathcal{P}(S^{2m-1})$ and

$$0 = c(x) = 2c(y) = \Sigma(y)$$

so that y has order 2, that is, $x = 0$. $\square$

Remark: [**Sah, 1981**] actually produced an inverse by the "Gauss-Bonnet" map $e : \mathcal{P}(S^n) \to \mathcal{P}(S^{n-1})$ given for a spherical simplex Δ by the "alternating angle sum"

$$e(\Delta) = \sum_F (-1)^{\dim F} [\theta(F, \Delta)]$$

where F runs through all non-empty faces of Δ (including Δ). Here $\theta(F, \Delta)$ is defined by choosing an interior point $x \in F$ and taking the $(n-1)$-dimensional spherical polytope formed by the closure of the set of all interior unit tangent vectors in Δ with origin at x.

EXAMPLE 7.8. $\mathcal{P}(S^2)$. By corollary 7.7 and example 7.2 we get an isomorphism $\mathcal{P}(S^2) \cong \mathcal{P}(S^1) \cong \mathbb{R}$ given by the area function.

For $\mathcal{P}(S^3)$ we now turn to the proof of the exact sequence c) in theorem 1.7. We use the same strategy as in example 7.2. In higher dimensions the exact sequence in theorem 3.12 gives rise to a "hyperhomology" spectral sequence (cf. Appendix A) converging to $H_*(O(n+1), \mathbb{Z}^t)$ with $E^1_{p,n+1} = H_p(O(n+1), \mathbb{Z})$ and for $q \leq n$:

$$
\begin{aligned}
E^1_{p,q} &= H_p\left(O(n+1), \left[\bigoplus_{U^q} \mathrm{Pt}(U^q)\right]^t\right) \\
&\cong H_p(O(q+1) \times O(n-q), \mathrm{Pt}(S^q)^t \otimes \mathbb{Z}^t) \\
&\cong \bigoplus_{i+j=p} H_i(O(q+1), \mathrm{Pt}(S^q)^t) \otimes H_j(O(n-q), \mathbb{Z}^t) \\
&\quad \bigoplus_{i+j=p-1} \mathrm{Tor}(H_i(O(q+1), \mathrm{Pt}(S^q)^t), H_j(O(n-q), \mathbb{Z}^t))
\end{aligned}
$$

(7.9)

by Shapiro's lemma and the Künneth theorem. We note the following easy facts:

PROPOSITION 7.10.   a) $E^r_{p,q} = 0$ *outside* $0 \leq q \leq n+1$.
b) *If* $q \not\equiv n \mod 2$ *then* $2E^r_{p,q} = 0$.
c)

$$
E^2_{0,q} = \begin{cases} \mathbb{Z}/2 & q = 0 \\ \mathcal{P}(S^n) & q = n \\ 0 & otherwise. \end{cases}
$$

d) $E^1_{p,0} \cong H_p(O(n), \mathbb{Z}^t)$.

Proof: a) is obvious.

b) follows from (7.9) and "center kills" for the element $(-\mathrm{id}_{n-q}) \in O(n-q)$.

c) By (7.9) we get for $0 < q < n$:

$$E^1_{0,q} = \mathcal{P}(S^q)/2\mathcal{P}(S^q) = 0$$

by Gerling's theorem. Since $E^1_{0,n+1} = \mathbb{Z}, E^1_{0,n} = \mathcal{P}(S^n)$ and since d^1 : $\mathbb{Z} \to \mathcal{P}(S^n)$ sends the generator to $[S^n]$ with non-zero volume we get

$$E^2_{0,n+1} = 0, \quad E^2_{0,n} = \mathcal{P}(S^n)/\mathbb{Z}.$$

d) follows from (7.9) and example 7.1.

EXAMPLE 7.11. $n = 3$. The E^1-term is as follows:

4	$\mathbb{Z}$			
3	$\mathcal{P}(S^3)$			
2	0	2-torsion	2-torsion	
1	0	$E^1_{1,1}$	$E^1_{2,1}$	
0	$\mathbb{Z}/2$	2-torsion	2-torsion	2-torsion
	0	1	2	3

where

(7.12,i) $E^1_{1,1} = H_1(O(2) \times O(2), \mathrm{Pt}(S^1)^t \otimes \mathbb{Z}^t)$

$\cong \mathcal{P}(S^1) \otimes H_1(O(2), \mathbb{Z}^t) \oplus H_1(O(2), \mathrm{Pt}(S^1)^t) \otimes \mathbb{Z}/2$

(7.12,ii) $E^1_{2,1} = H_2(O(2) \times O(2), \mathrm{Pt}(S^1)^t \otimes \mathbb{Z}^t)$.

In (7.12,i) the first term is isomorphic to $\mathbb{R} \otimes (\mathbb{R}/\mathbb{Z})$ and the second term has at most order 2 by example 7.2. Furthermore $d^2 : \mathcal{P}(S^n)/\mathbb{Z} \to \mathbb{R} \otimes (\mathbb{R}/\mathbb{Z})$ is identified with the Dehn-invariant. Similarly one shows in (7.12,ii) that $E^1_{2,1}$ is annihilating by 2. Hence we conclude the following:

PROPOSITION 7.13. *There is an exact sequence*

$$H_3(O(4), \mathbb{Z}^t) \to \mathcal{P}(S^3)/\mathbb{Z} \xrightarrow{D} \mathbb{R} \otimes \mathbb{R}/\mathbb{Z} \to H_2(O(4), \mathbb{Z}^t)$$

where the kernel of the first map is annihilated by 8 and the cokernel of the last map is annihilated by 4.

Remark: Since the double covering group Pin(4) of O(4) has the form Pin(4) $= (S_1 \times S_2) \rtimes (\mathbb{Z}/2)$ with $S_i \cong \mathrm{Sp}(1) \cong \mathrm{SU}(2)$ as explained in chapter 6, one conclude that

$$H_i(O(4), \mathbb{Z}^t) \cong H_i(\mathrm{SU}(2), \mathbb{Z}) \oplus \text{ 2-torsion, for } i \leq 3.$$

Hence we have established the exact sequence c) in theorem 1.7 except for the kernel of the beginning map and the cokernel of the end map, which may à priori be groups annihilated by 8 respectively 4. One can get rid of these 2-torsion groups by a closer analysis of the above spectral sequence using the group Pin(4) all the way through. Alternatively one can use a different description of $\mathcal{P}(S^n)/\Sigma\mathcal{P}(S^{n-1})$ as we shall now indicate.

As in chapter 3 we let $\bar{C}_*(S^n)$ denote the chain complex of all tuples of points in S^n. Then one can prove the following:

THEOREM 7.14. *For $n > 0$ there is a natural isomorphism*
$$\mathcal{P}(S^n)/\Sigma\mathcal{P}(S^{n-1}) \cong H_n(SO(n+1)\backslash\bar{C}_*(S^n)).$$

Proof: By Gerling's theorem $\mathcal{P}(S^n) = \mathcal{P}(S^n, SO(n+1))$ and as in the proof of theorem 7.4 we obtain an exact sequence

$$\mathcal{P}(S^{n-1}) = H_0(S(O(n) \times O(1)), \mathrm{Pt}(S^{n-1})) \xrightarrow{\Sigma} H_0(SO(n+1), \mathrm{Pt}(S^n))$$
$$\xrightarrow{h} H_0(SO(n+1), \mathrm{St}(S^n)) \to 0,$$

where $S(O(n) \times O(1)) = SO(n+1) \cap (O(n) \times O(1))$. That is,

$$\mathcal{P}(S^n)/\Sigma\mathcal{P}(S^{n-1}) \cong H_0(SO(n+1), H_n(\bar{C}(S^n)/\bar{C}_*(S^n)^{n-1}))$$
$$\cong H_n\left(\frac{\bar{C}_*(S^n)}{SO(n+1)} \Big/ \frac{\bar{C}_*(S^n)^{n-1}}{SO(n+1)}\right)$$

and we claim that the map

$$H_n\left(\frac{\bar{C}_*(S^n)}{SO(n+1)}\right) \to H_n\left(\frac{\bar{C}_*(S^n)}{SO(n+1)} \Big/ \frac{\bar{C}_*(S^n)^{n-1}}{SO(n+1)}\right)$$
$$\cong \mathcal{P}(S^n)/\Sigma\mathcal{P}(S^{n-1})$$

is an isomorphism. An inverse is provided by the composite map in the diagram

$$H_0(O(n+1), \mathrm{St}(S^n)^t) \xrightarrow{\simeq} H_n\left(\frac{\bar{C}_*(S^n)^t}{O(n+1)} \Big/ \frac{\bar{C}_*(S^n)^{n-1}}{O(n+1)}\right)$$

$$h \Big\uparrow \qquad\qquad\qquad\qquad \Big\downarrow$$

$$\mathcal{P}(S^n)/\Sigma\mathcal{P}(S^{n-1}) \longrightarrow H_n\left(\frac{\bar{C}_*(S^n)}{SO(n+1)}\right)$$

where the right hand vertical map is induced by sending a chain c to $c - c'$ with c' the mirror image of c. We refer to [**Dupont-Parry-Sah, 1988**, §5] for the details.   □

Remark 1.   The hyperhomology spectral sequence for $\mathrm{Spin}(4) = S_1 \times S_2$ acting on $\bar{C}_*(S^3)$ gives similarly to proposition 7.13 the following exact sequence (see again [**Dupont-Parry-Sah, 1988**, §5] for details):

(7.15)
$$0 \to \mathbb{Q}/\mathbb{Z} \to H_3(\mathrm{SU}(2)) \to \mathcal{P}(S^3)/\Sigma\mathcal{P}(S^2) \xrightarrow{d^2} \mathbb{R}/\mathbb{Z} \otimes \mathbb{R}/\mathbb{Z} \to$$
$$\to H_2(\mathrm{SU}(2)) \to 0$$

where $\mathbb{Q}/\mathbb{Z}$ is the 3rd homology group of the subgroup of $\mathrm{U}(1) \subseteq \mathrm{Sp}(1) = \mathrm{SU}(2)$ consisting of the roots of unity. Comparing this with the sequence in proposition 7.13 gives the exactness at the beginning and end of the sequence c) in theorem 1.7.

Remark 2.   A *lune* in S^n of angle $\theta \in [0, 2\pi]$ is a polytope of the form

$$L(\theta) = \text{ closure } \{x \in S^n \mid (x_1, x_2) = (r\cos u, r\sin u), \quad r > 0, 0 \leq u \leq \theta\}$$

i.e. the join of an arc of length θ with S^{n-2}. $L(\theta)$ is called a *rational* lune if $\theta/2\pi$ is rational. Clearly $\mathrm{Vol}(L(\theta))/\mathrm{Vol}(S^n) = \theta/2\pi$ so in particular for $L(\theta)$ a rational lune $[L(\theta)]$ represents $\theta/2\pi \in \mathbb{Q} \subseteq \mathcal{P}(S^n)$. Note that $L(2\pi/q)$ is a fundamental domain for the cyclic group of order q generated by the rotation of angle $2\pi/q$ in the (x_1, x_2)-plane. It is shown in [**Dupont-Sah, 1982**] that any spherical polytope which is the fundamental domain for a finite subgroup G of $\mathrm{O}(n+1)$ is scissors congruent to a rational lune i.e. to $L(2\pi/|G|)$.

Remark 3.   In the spectral sequence (7.9) there is an edge homomorphism

$$\sigma : H_n(\mathrm{O}(n+1), \mathbb{Z}^t) \to E_{0,n}^\infty \subseteq E_{0,n}^2 = \mathcal{P}(S^n)/\mathbb{Z}.$$

To describe this explicitly on the chain level we consider both the bar complex $B_*(\mathrm{O}(n+1))$ and the chain complex $C_*(S^n)$ as chain complexes of $\mathbb{Z}[\mathrm{O}(n+1)]$-modules augmented to the trivial $\mathrm{O}(n+1)$-module $\mathbb{Z}$ and since $C_*(S^n)$ is $(n-1)$-connected we can extend the identity $\mathbb{Z} \to \mathbb{Z}$ to a chain map of $\mathrm{O}(n+1)$-modules

$$\hat{\sigma} : B_i(\mathrm{O}(n+1)) \to C_i(S^n) \qquad \text{for } i \leqq n.$$

After tensoring with the $O(n+1)$-module $\mathbb{Z}^t$ we get the desired map on the chain level

$$B_n(O(n+1)) \otimes_{\mathbb{Z}[O(n+1)]} \mathbb{Z}^t \xrightarrow{\quad \hat{\sigma} \quad} C_n(S^n) \otimes_{\mathbb{Z}[O(n+1)]} \mathbb{Z}^t$$

$$\searrow {\scriptstyle \sigma} \qquad\qquad \downarrow$$

$$[C_n(S^n)/C_n(S^n)^{n-1}] \otimes_{\mathbb{Z}[O(n+1)]} \mathbb{Z}^t.$$

For the case of the spectral sequence associated to the sequence in theorem 3.11

$$E^1_{p,q} = H_p(O(q+1) \times O(n-q), \mathrm{St}(S^q)^t \otimes \mathbb{Z}^t) \Rightarrow H_{p+q}(O(n+1), \mathbb{Z}^t)$$

the corresponding edge map

$$\bar{\sigma} : H_n(O(n+1), \mathbb{Z}^t) \to E^1_{0,n} = \mathcal{P}(S^n)/\Sigma\mathcal{P}(S^{n-1})$$

$$\cong H_n\left([\bar{C}_*(S^n)/\bar{C}_*(S^n)^{n-1}] \bigotimes_{\mathbb{Z}[O(n+1)]} \mathbb{Z}^t \right)$$

is given on the chain level by choosing $x_0 \in S^n$ and putting

$$\bar{\sigma}[g_1|\ldots|g_n] = (x_0, g_1 x_0, g_1 g_2 x_0, \ldots, g_1 \ldots g_n x_0).$$

We shall return to this in chapter 10.

CHAPTER 8

Hyperbolic scissors congruence

We now turn to the case of hyperbolic geometry. In particular in this chapter we shall obtain the exact sequence in theorem 1.7, b). It is convenient to use various models for the hyperbolic n-space $\mathcal{H}^n$ depending on the situation and we shall once and for all refer to [**Iversen, 1992**] for the identification between them. In chapter 2 we considered the "quadric" model but we shall also make use of the "disk" model

$$D^n = \{x = (x_1, \ldots, x_n) \mid |x| < 1\}$$

or the "upper half space" model

$$\mathbb{R}^n_+ = \{x = (x_1, \ldots, x_n) \mid x_n > 0\}$$

which are the Riemannian spaces with the metric given by

$$ds^2 = 4(1 - |x|^2)^{-1} \sum_i dx_i^2 \quad \text{for } D^n,$$

$$ds^2 = (1/x_n^2) \sum_i dx_i^2 \quad \text{for } \mathbb{R}^n_+.$$

For the study of polytopes it is necessary to know the geodesic subspaces in each model: In the "quadric" they are just the non-empty intersections with linear subspaces. In the "disk" or "upper half space" they are the non-empty intersections with spheres or hyperplanes intersecting the "boundary" at $90°$. The set of boundary points $\partial \mathcal{H}^n$ is in the case of D^n visibly the $(n-1)$-sphere S^{n-1} and in the case of $\mathbb{R}^n_+$, $\partial \mathcal{H}^n = (\mathbb{R}^{n-1} \times 0) \cup \{\infty\}$, and points on the boundary are usually called "ideal points". We shall also use the notation $\bar{\mathcal{H}} = \mathcal{H}^n \cup \partial \mathcal{H}^n$.

Also let us denote the group of isometries for $\mathcal{H}^n$ by $G(n)$. As mentioned in chapter 2, $G(n) \cong O^1(1, n)$.

Let us now try to compute the scissors congruence groups $\mathcal{P}(\mathcal{H}^n)$. For $n = 1, \mathcal{H}^1$ and E^1 are isometric via the exponential map; hence $\mathcal{P}(\mathcal{H}^1) \cong \mathbb{R}$ and again the isomorphism is given by the "length" function. For $n > 1$ we could of course proceed as in the previous chapter and try to calculate the hyperhomology spectral sequence for the Lusztig exact sequence in theorem 3.11. However it is convenient to make use of the geometry of the boundary $\partial \mathcal{H}^n$ as follows:

First notice that it makes good sense to consider polytopes where some of the vertices are allowed to be ideal points and we get in this way a scissors congruence group $\mathcal{P}(\bar{\mathcal{H}}^n)$. Note however that Zylev's theorem (cf. the remarks following definition 1.6) that stable scissors congruence implies scissors congruence, does *not* hold in this case:

EXAMPLE 8.1. In the upper half plane model for $\mathcal{H}^2$ consider the triangle $\triangle$ bounded by two vertical lines going to the vertex ∞ and a half circle perpendicular to the boundary (see figure). If we bisect the strip between the lines and draw the half circles intersecting each piece at the same angles as in $\triangle$ we obtain a scissors congruence

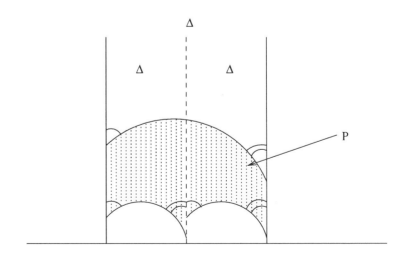

$$\triangle \amalg \triangle \simeq \triangle \amalg P$$

where P is a polygon with only finite vertices. Hence in the scissors congruence group $[\Delta] = [P]$ but since Δ has one ideal vertex Δ and P cannot be scissors congruent.

We would also like to define a scissors congruence group $\mathcal{P}(\partial \mathcal{H}^n)$ where *only* ideal vertices are allowed; but here it is less clear how to define scissors congruence since cutting with a hyperplane invariably introduces finite vertices. However notice that theorem 2.10 remains valid also for $X = \bar{\mathcal{H}}^n$ so that (given an orientation of $\bar{\mathcal{H}}^n$ - e.g. the canonical one) we get an isomorphism

$$(8.2) \qquad \mathcal{P}(\bar{\mathcal{H}}^n) \cong H_0(G(n), H_n(\bar{C}_*(\bar{\mathcal{H}}^n)/\bar{C}_*(\bar{\mathcal{H}}^n)^{n-1})^t)$$

where $\bar{C}_*(\bar{\mathcal{H}}^n)$ denotes the chain complex of all tuples of points in $\bar{\mathcal{H}}^n$ with the rank filtration $\bar{C}_*(\bar{\mathcal{H}}^n)^p, p = 0, 1, 2, \ldots$. We can therefore simply define

$$(8.3) \qquad \mathcal{P}(\partial \mathcal{H}^n) = H_0(G(n), H_n(\bar{C}_*(\partial \mathcal{H}^n)/\bar{C}_*(\partial \mathcal{H}^n)^{n-1})^t)$$

where $\bar{C}_*(\partial \mathcal{H}^n)$ are defined in the obvious way but only allowing points on $\partial \mathcal{H}^n$.

Explicitly $\mathcal{P}(\partial \mathcal{H}^n)$ is the abelian group with generators $(a_0, \ldots, a_n)$ with $a_i \in \partial \mathcal{H}^n$, and relations:

$(8.4,\mathrm{i}) \qquad (a_0, \ldots, a_n) = 0 \quad$ if $\quad \{a_0, \ldots, a_n\}$ lies on a hyperplane,

$(8.4,\mathrm{ii}) \quad \displaystyle\sum_{i=0}^{n+1} (-1)^i (a_0, \ldots, \hat{a}_i, \ldots, a_n) = 0 \quad$ for $\quad a_0, \ldots, a_{n+1} \in \partial \mathcal{H}^n$

$(8.4,\mathrm{iii}) \qquad\qquad (ga_0, \ldots, ga_n) = \det(g)(a_0, \ldots, a_n) \quad g \in G(n).$

We now have the following:

THEOREM 8.5.   a) *The inclusion $\mathcal{H}^n \subset \bar{\mathcal{H}}^n$ induces an isomorphism $\iota_n : \mathcal{P}(\mathcal{H}^n) \xrightarrow{\cong} \mathcal{P}(\bar{\mathcal{H}}^n)$ for $n > 1$.*

   b) *There is an exact sequence*

$$H_1(O(n), \mathrm{St}(S^{n-1})^t) \to \mathcal{P}(\partial \mathcal{H}^n) \to \mathcal{P}(\bar{\mathcal{H}}^n) \to H_0(O(n), \mathrm{St}(S^{n-1})^t) \to 0$$

   c) *In particular for n odd $\mathcal{P}(\partial \mathcal{H}^n) \to \mathcal{P}(\bar{\mathcal{H}}^n)$ is surjective with kernel consisting of elements of order at most 2.*

   d) *Also for $n > 0$ even there is an isomorphism*

$$\mathcal{P}(\bar{\mathcal{H}}^n)/\mathrm{im}(\mathcal{P}(\partial \mathcal{H}^n) \to \mathcal{P}(\bar{\mathcal{H}}^n)) \cong \mathcal{P}(S^{n-1})/\Sigma \mathcal{P}(S^{n-2}).$$

Sketch proof. c) and d) clearly follows from b) together with "center kills" and theorem 7.4. Let us sketch the proof of a) - the proof of b) is similar. Similarly to the isomorphism in theorem 3.5

$$(8.6) \quad \mathcal{P}(\mathcal{H}^n) \cong H_0(G(n), \mathrm{St}(\mathcal{H}^n)^t), \quad \mathrm{St}(\mathcal{H}^n) = \tilde{H}_{n-1}(\mathcal{T}(\mathcal{H}^n), \mathbb{Z})$$

we have

$$(8.7) \quad \mathcal{P}(\bar{\mathcal{H}}^n) \cong H_0(G(n), \mathrm{St}(\bar{\mathcal{H}}^n)^t), \quad \mathrm{St}(\bar{\mathcal{H}}^n) = \tilde{H}_{n-1}(\mathcal{T}(\bar{\mathcal{H}}^n), \mathbb{Z}),$$

where ideal points are allowed in the Tits complex for $\mathcal{T}(\bar{\mathcal{H}}^n)$. Also for $p \in \partial \mathcal{H}^n$ we have a Tits complex $\mathcal{T}(\bar{\mathcal{H}}^n, p)$ of flags ending in p, and we consider

$$\mathrm{St}(\bar{\mathcal{H}}^n, p) = \tilde{H}_{n-2}(\mathcal{T}(\bar{\mathcal{H}}^n, p), \mathbb{Z})$$

as a G_p-module, where $G_p \subseteq G(n)$ is the isotropy subgroup at p. Then one shows that there is an exact sequence of $G(n)$-modules

$$0 \to \mathrm{St}(\mathcal{H}^n) \to \mathrm{St}(\bar{\mathcal{H}}^n) \to \bigoplus_{p \in \partial \mathcal{H}^n} \mathrm{St}(\bar{\mathcal{H}}^n, p) \to 0$$

from which we obtain the exact sequence

$$(8.8) \quad \begin{aligned} \cdots &\to H_1\left(G(n), \left[\bigoplus_{p \in \partial \mathcal{H}^n} \mathrm{St}(\bar{\mathcal{H}}^n, p)\right]^t\right) \mathcal{P}(\mathcal{H}^n) \xrightarrow{i_n} \mathcal{P}(\bar{\mathcal{H}}^n) \to \\ &\to H_0\left(G(n), \left[\bigoplus_{p \in \partial \mathcal{H}} \mathrm{St}(\bar{\mathcal{H}}^n, p)\right]^t\right) \end{aligned}$$

Here by Shapiro's lemma

$$H_i\left(G(n), \left[\bigoplus_{p \in \partial \mathcal{H}^n} \mathrm{St}(\bar{\mathcal{H}}^n, p)\right]^t\right) \cong H_i(G_\infty, \mathrm{St}(\bar{\mathcal{H}}^n, \infty)^t)$$

and in the upper half space model there is a 1-1 correspondence between the geodesic subspaces through ∞ and the affine subspaces of the boundary $\mathbb{R}^{n-1}$. Hence $\mathrm{St}(\bar{\mathcal{H}}^n, \infty) \cong \mathrm{St}(E^{n-1})$ and via this isomorphism G_∞ acts as the group of *similarities*

$$G_\infty \cong \mathrm{Sim}(n-1) = T(n-1) \ltimes \mathrm{Sim}_0(n-1), \quad \mathrm{Sim}_0(n-1) = \mathrm{O}(n-1) \times \mathbb{R}_+^\times$$

Hence a) follows from the sequence (8.8) and the following claim:

$$(8.9) \quad H_*(\mathrm{Sim}(n), \mathrm{St}(E^n)^t) = 0 \quad \text{for } n > 0.$$

To show (8.9) one uses the Hochschild-Serre spectral sequence of the extension
$$1 \to T(n) \to \mathrm{Sim}(n) \to \mathrm{Sim}_0(n) \to 1$$
with E^2-term $H_*(\mathrm{Sim}_0(n), H_*(T(n), \mathrm{St}(E^n))^t)$. However by the argument of chapter 4, $H_*(T(n), \mathrm{St}(E^n))^t)$ consists of rational vector spaces on which the dilatation μ_a by a positive integer a acts by multiplication by a^q for $q > 0$. Hence by "center kills" the E^2-term vanishes which proves (8.9), and ends the proof of a). The proof of b) is similar (but easier) and we refer to [**Dupont-Sah, 1982**] for the details.   □

Now let us return to $\mathcal{P}(\mathcal{H}^n)$ in low dimensions:

EXAMPLE 8.10. $n = 2$: By theorem 8.5 we have $\mathcal{P}(\mathcal{H}^2) \simeq \mathcal{P}(\bar{\mathcal{H}}^2)$ and there is an exact sequence
(8.11)
$$H_1(O(2), \mathrm{St}(S^1)^t) \to \mathcal{P}(\partial\mathcal{H}^2) \to \mathcal{P}(\bar{\mathcal{H}}^2) \to \mathcal{P}(S^1)/\Sigma\mathcal{P}(S^0) \to 0.$$
Now in the upper half plane model $\partial\mathcal{H}^2 = \mathbb{R} \cup \{\infty\}$ and $G(2) \cong \mathrm{PSl}(2, \mathbb{R}) \ltimes (\mathbb{Z}/2)$, where $g \in \mathrm{PSl}(2, \mathbb{R})$ acts by

$$g(z) = (az + b)/(cz + d) \quad \text{for} \quad z \in \mathbb{R} \cup \{\infty\}, \quad g = \begin{pmatrix} a & b \\ c & d \end{pmatrix},$$

and $\mathbb{Z}/2$ by multiplication by ± 1. Since $\mathcal{P}(\partial\mathcal{H}^2)$ is generated by all triples of points on $\partial\mathcal{H}^2$ and since $\mathrm{PSl}(2, \mathbb{R})$ acts transitively on such triples it follows from (8.4) that $\mathcal{P}(\partial\mathcal{H}^2) \cong \mathbb{Z}$. By example 7.2 the first term in (8.11) has order at most 2 and $\mathcal{P}(S^1)/\Sigma\mathcal{P}(S^0) \cong \mathbb{R}/\mathbb{Z}\pi$; hence we obtain that
$$0 \to \mathbb{Z} \to \mathcal{P}(\bar{\mathcal{H}}^2) \to \mathbb{R}/\mathbb{Z}\pi \to 0$$
is exact. From this it clearly follows that the "Area"-function provides an isomorphism $\mathcal{P}(\bar{\mathcal{H}}^2) \cong \mathbb{R}$.

For $n = 3$ theorem 8.5 gives that $\mathcal{P}(\mathcal{H}^3) \cong \mathcal{P}(\bar{\mathcal{H}}^3)$ and the inclusion $\partial\mathcal{H}^3 \subset \bar{\mathcal{H}}^3$ induces a surjection $\mathcal{P}(\partial\mathcal{H}^3) \to \mathcal{P}(\bar{\mathcal{H}}^3)$ with kernel consisting of elements of order 2. As we shall see later $\mathcal{P}(\partial\mathcal{H}^3)$ is actually torsion free so that indeed $\mathcal{P}(\partial\mathcal{H}^3) \cong \mathcal{P}(\bar{\mathcal{H}}^3) \cong \mathcal{P}(\mathcal{H}^3)$. In the upper half space model we identify the boundary with the Riemann sphere $\partial\mathcal{H}^3 = \mathbb{C} \cup \{\infty\}$ and $G(3) = \mathrm{PSl}(2, \mathbb{C}) \ltimes \mathbb{Z}/2$ acts by

$$g(z) = \frac{az + b}{cz + d} \quad \text{for } g = \begin{pmatrix} a & b \\ c & c \end{pmatrix} \in \mathrm{Sl}(2, \mathbb{C})$$

and the generator of $\mathbb{Z}/2$ acts by $z \mapsto -\bar{z}$. Now it is convenient to consider coinvariance with respect to the subgroup $\mathrm{PSl}(2,\mathbb{C})$, that is, we define an abelian group $\mathcal{P}_{\mathbb{C}}$ with generators $(a_0, \ldots, a_3)$, $(a_i \in \mathbb{C} \cup \{\infty\})$ modulo the relations

(8.12,i) $(a_0, \ldots, a_3) = 0$ if $a_i = a_j$ for some $i \neq j$.

(8.12,ii)

$$\sum_{i=0}^{4} (-1^i (a_0, \ldots, \hat{a}_i, \ldots, a_4) = 0 \text{ for all 5-tuples of distinct } a_i\text{'s.}$$

(8.12,iii) $(a_0, \ldots, a_3) = (ga_0, \ldots, ga_3)$ for $g \in \mathrm{PSl}(2,\mathbb{C})$.

Now, up to the action of $\mathrm{PSl}(2,\mathbb{C})$, each $(a_0, \ldots, a_4)$ is determined by the cross-ratio

$$z = \{a_0 : \cdots : a_4\} = \frac{a_0 - a_2}{a_0 - a_3} \frac{a_1 - a_3}{a_1 - a_2} \in \mathbb{C} - \{0, 1\}.$$

Indeed, for this z,

$$(a_0, \ldots, a_3) = g(\infty, 0, 1, z) \quad \text{for some } g \in \mathrm{PSl}(2,\mathbb{C}).$$

With this motivation we introduce

Definition: Let F be any field. Then $\mathcal{P}_F$ is the abelian group with generators $\{z\}, z \in F - \{0, 1\}$ and defining "five-term" relations:

(8.13)
$$\{z_1\} - \{z_2\} + \{z_2/z_1\} - \{(1 - z_2)/(1 - z_1)\} + \{(1 - z_2^{-1})/(1 - z_1^{-1})\} = 0$$
$$z_1, z_2 \in \mathbb{C} - \{0, 1\}, z_1 \neq z_2.$$

Remark: This relation is just a reformulation of (8.12,ii). When F is algebraically closed one deduces the following relations for $z \in F - \{0, 1\}$ (cf. [**Dupont-Sah, 1982**, lemma 5.11])

(8.13,ii) $\{z\} + \{z^{-1}\} = 0$

(8.13,iii) $\{z\} + \{1 - z\} = 0.$

We can then introduce the symbols $\{0\} = \{1\} = \{\infty\} = 0$ and (8.13), i)-iii) remain valid for $z_i \in F \cup \{\infty\}$.

When F is algebraically closed one can prove a useful generalization of (8.13) known as "Rogers' identity" which we shall now describe: Let

$f, g \in F(t)$, the field of rational functions and write

$$f(t) = a \prod_i (\alpha_i - t)^{d(i)}, \quad g(t) = b \prod_j (\beta_j - t)^{e(j)}$$

where $d(i), e(j) \in \mathbb{Z}$ and $\alpha_i \in F$ distinct and $\beta_j \in F$ distinct. Put

$$f^- * g = \sum_{i,j} d(i) e(j) \{\alpha_i^{-1} \beta_j\} \in \mathcal{P}_F$$

where the sum extends over i, j with $\alpha_i, \beta_j \in F - \{0\}$ and the expression is 0 if $f, g \in F$. We refer to [**Dupont-Sah, 1982**] for the proof of the following:

THEOREM 8.14. *Let F be algebraically closed and let $f \in F(t)$. Then the following holds in $\mathcal{P}_F$:*

$$f^- * (1 - f) = \{f(0)\} - \{f(\infty)\}.$$

Remark: For $f(t) = c(\alpha - t)/(\beta - t)$ one checks that (8.14) is equivalent to (8.13) with $z_1 = c, z_2 = c\alpha/\beta$.

COROLLARY 8.15 ("The distribution identity"). *Let F be algebraically closed of characteristic 0. Let $n \in \mathbb{N}$ and $\zeta \in F$ a primitive root of unity. Then for $z \in F$ arbitrary the following holds in $\mathcal{P}_F$:*

$$\{z^n\} = n \sum_{j=0}^{n-1} \{\zeta^i z\}.$$

Proof: For $f(t) = (1 - t^n)/(1 - z^n) = \left(\prod_{i=0}^{n-1} (\zeta^i - t) \right) / (1 - z^n)$ we have

$$1 - f(t) = (t^n - z^n)/(1 - z^n) = \left(\prod_{j=0}^{n-1} (z\zeta^j - t) \right) / (z^n - 1),$$

and (8.15) follows from (8.14), since

$$\{f(0)\} - \{f(\infty)\} = \{(1 - z^n)^{-1}\} - \{\infty\} = \{z^n\}.$$

$\square$

From this one deduces (cf. [**Suslin, 1986**], [**Suslin, 1991**])

THEOREM 8.16. *For F algebraically closed of characteristic zero $\mathcal{P}_F$ is uniquely divisible.*

Sketch proof.   That $\mathcal{P}_F$ is divisible is obvious from corollary 8.15. To prove unique divisibility one must prove that for $n \in \mathbb{N}$, and ζ a primitive n'th root of 1,

$$\{w\}/n = \sum_{j=0}^{n-1}\{\zeta^j w^{1/n}\}$$

is well-defined, i.e. respects the relation (8.13). In general this is rather complicated and uses more algebraic geometry. We refer to [**Suslin, 1986**] for the details. For $n = 2$ however the proof is rather simple: We must show that $\{z\}/2 = \{z^{\frac{1}{2}}\} + \{-z^{\frac{1}{2}}\}$ satisfy the identity

(8.17) $\qquad \begin{aligned} \{z_1\}/2 - \{z_2\}/2 + \{z_2/z_1\}/2 - \{(z_2 - 1)/(z_1 - 1)\}/2 + \\ + \{(z_2^{-1} - 1)/(z_1^{-1} - 1)\}/2 = 0 \end{aligned}$

for $z_1, z_2 \in F - \{0, 1\}, z_1 \neq z_2$. For this we write $z_i = w_i^2, i = 1, 2$, and consider

$$f(t) = -w_1 \frac{(t - \alpha_1)(t - \alpha_2)}{(t - 1)(t + \alpha_2)}, \quad \alpha_1 = \frac{w_2}{w_1}, \alpha_2 = \frac{1 + w_2}{1 - w_1}.$$

Direct computation shows that $1 - f(t)$ has simple zeroes at $\pm(w_2^2 - 1)^{\frac{1}{2}}/(w_1^2 - 1)^{\frac{1}{2}}$ and simple poles at 1 and $-\alpha_2$. Using the fact that $\{-1\} = 2(\{i\} + \{-i\}) = 0$ in $\mathcal{P}_F$ we conclude from theorem 8.14:

$$\begin{aligned} \{w_2\} - \{-w_1\} = \{(z_2^{-1} - 1)/(z_1^{-1})\}/2 - \{(z_2 - 1)/(z_1 - 1)\}/2 - \\ - \{w_1/w_2\} - \{-(w_1 - 1)/(w_2 - 1)\} - \{-(w_2^{-1} + 1)/(w_1^{-1} - 1)\}. \end{aligned}$$

Now replacing z_1, z_2 by $-w_2, w_1$ in (8.13) and adding the result to the above equation gives (8.17). $\qquad \square$

COROLLARY 8.18. *There are canonical isomorphisms of rational vector spaces*

$$\mathcal{P}(\mathcal{H}^3) \cong \mathcal{P}(\bar{\mathcal{H}}^3) \cong \mathcal{P}(\partial\mathcal{H}^3) \cong \mathcal{P}_{\mathbb{C}}^-$$

where $\mathcal{P}_{\mathbb{C}}^-$ is the (-1)-eigenspace for $\{z\} \mapsto \{\bar{z}\}$.

Proof: Comparing (8.4) and (8.12) we get

$$\mathcal{P}(\partial\mathcal{H}^3) \cong \mathcal{P}_{\mathbb{C}}/\operatorname{span}\{\{z\} + \{\bar{z}\}, \{r\} \mid z \in \mathbb{C}, r \in \mathbb{R}\}.$$

However

$$\{r\} = 2\left(\{r^{\frac{1}{2}}\} + \{-r^{\frac{1}{2}}\}\right)$$

so that the last group equals $\mathcal{P}_{\mathbb{C}}/\operatorname{span}\{\{z\} + \{\bar{z}\} \mid z \in \mathbb{C}\} \cong \mathcal{P}_{\mathbb{C}}^-$ since $\mathcal{P}_{\mathbb{C}}$ is 2-divisible. This proves the last isomorphism in (8.18). Since $\mathcal{P}(\partial\mathcal{H}^3) \to \mathcal{P}(\bar{\mathcal{H}}^3)$ is surjective with kernel annihilated by 2 and since $\mathcal{P}(\partial\mathcal{H}^3)$ is torsion free the middle isomorphism is proved. □

For $\mathcal{P}_{\mathbb{C}}$ there is an exact sequence similar to b) in theorem 1.7. This is a reformulation of a result of D. Wigner and S. Bloch (unpublished, cf. [**Bloch, 1978**]):

THEOREM 8.19. *Let F be algebraically closed of characteristic 0. Then there is an exact sequence*

$$0 \to \mathbb{Q}/\mathbb{Z} \to H_3(\mathrm{Sl}(2, F)) \xrightarrow{\sigma} \mathcal{P}_F \xrightarrow{\lambda} \wedge_{\mathbb{Z}}^2(F^\times/\mu_F) \to H_2(\mathrm{Sl}(2, F)) \to 0.$$

Here $\mathbb{Q}/\mathbb{Z} = H_3(\mu_F, \mathbb{Z})$ where $\mu_F \subseteq F^\times$ is the group of roots of unity and the first and last map is induced by $\zeta \mapsto \begin{pmatrix} \zeta & 0 \\ 0 & \zeta^{-1} \end{pmatrix}$. The map σ is given by $\sigma([g_1|g_2|g_3]) = \{\infty : g_1\infty : g_1 g_2 \infty : g_1 g_2 g_3 \infty\} \in \mathcal{P}_{\mathbb{C}}$. Also λ is given by $\lambda(z) = z \wedge (1 - z)$ and the last map sends $a \wedge b$ to $[a|b] - [b|a], a, b \in F^\times \subseteq \mathrm{Sl}(2, F)$.

Proof: We consider the action by $G = \mathrm{Sl}(2, F)$ on $P^1(F) = F \cup \{\infty\}$ as before which induces an action on the chain complex $C_* = C_*(P^1(F))$ of $(k + 1)$-tuples in degree k, $(a_0, \ldots, a_k), a_i \in P^1(F)$, of *distinct* points. Since F is infinite it is easy to see similarly to lemma 3.6 that the chain complex is acyclic, and since $G/\{\pm 1\}$ is exactly 3-transitive on $P^1(F)$ we have that C_k is induced from $\{\pm 1\} \subseteq G$, for $k \geq 2$. Hence for $k \geq 2$ we have

$$H_i(G, C_k) = \begin{cases} \mathbb{Z}[G'^{k-2}], & i = 0, \\ \mathbb{Z}/2[G'^{k-2}], & i \text{ odd} \\ 0 & i \text{ even} > 0. \end{cases}$$

Furthermore for $k = 0$ and 1 the isotropy subgroup is

$$B = \left\{ \left(\begin{array}{cc} a & b \\ 0 & a^{-1} \end{array} \right) \mid a \neq 0 \right\} \quad \text{for } k = 0,$$

$$T = \left\{ \left(\begin{array}{cc} a & 0 \\ 0 & a^{-1} \end{array} \right) \mid a \neq 0 \right\} \quad \text{for } k = 1.$$

Hence, by Shapiro's lemma, "center kills" and proposition 4.7,

$$H_i(G, C_0) \cong H_i(G, C_1) \cong H_i(F^\times, \mathbb{Z}) \cong \wedge_{\mathbb{Z}}^i (F^\times / \mu_F) \oplus H_i(\mu_F, \mathbb{Z}).$$

The hyperhomology spectral sequence (Appendix A)

$$E_{p,q}^1 = H_p(G, C_q) \Rightarrow H_{p+q}(G)$$

thus has the following E^1-term

4	$H_0(G, C_4)$			
3	$H_0(G, C_3)$	2-torsion	0	2-torsion
2	$\mathbb{Z}$	$\mathbb{Z}/2$	0	$\mathbb{Z}/2$
1	$\mathbb{Z}$	$F^\times$	$\wedge^2(F^\times/\mu_F)$	$\wedge^3(F^\times/\mu_F) \oplus \mathbb{Q}/\mathbb{Z}$
0	$\mathbb{Z}$	$F^\times$	$\wedge^2(F^\times/\mu_F)$	$\wedge^3(F^\times/\mu_F) \oplus \mathbb{Q}/\mathbb{Z}$
	0	1	2	3

with generator for $q = 0, 1, 2$ being respectively (∞), $(\infty, 0)$ and $(\infty, 0, 1)$. Since $w = \left(\begin{array}{cc} 0 & 1 \\ -1 & 0 \end{array} \right)$ induces $z \mapsto -z^{-1}$ in $F^\times$ and interchanges ∞ and 0 it is easy to compute d^1 and we get for E^2:

3	$\mathcal{P}_{\mathbb{C}}$	2-torsion	0	
2	0	0	0	
1	0	0	$\wedge^2(F^\times/\mu_F)$	$\mathbb{Q}/\mathbb{Z}$
0	$\mathbb{Z}$	0	$\wedge^2(F^\times/\mu_F)$	$\mathbb{Q}/\mathbb{Z}$
	0	1	2	3

Here we claim that $E_{2,1}^\infty = 0$, i.e. that the natural map

$$\wedge^2(F^\times/\mu_F) = E_{2,1}^2 \to E_{2,1}^\infty \to H_3(G, \mathbb{Z})/(\mathbb{Q}/\mathbb{Z})$$

vanishes. To see this we split the chain complex C_* into exact sequences,

$$0 \to Z_0 \to C_0 \to \mathbb{Z} \to 0$$
$$0 \to Z_1 \to C_1 \to Z_0 \to 0$$

etc., and the map above is the composite

$$E_{2,1}^2 = H_2(G, C_1) \to H_2(G, Z_0) \stackrel{\partial^{-1}}{\to} H_3(G, \mathbb{Z})/H_3(G, C_0)$$

But the first map sends the generator $(\infty, 0)$ to $(0) - (\infty)$ in Z_0 and the element w above acts as (-1) on this but induces the identity on $H_2(T) = \wedge^2(F^\times/\mu)$. Hence the image is annihilated by 2 and is therefore 0.

The exact sequence in theorem 8.19 now easily follows from the above E^2-term. It is straight forward but somewhat tedious to identify the maps in the sequence (actually strictly speaking λ should be multiplied by 2) and we just refer to [**Dupont-Sah, 1982**] for the details.   □

Remark 1.   We now obtain the exact sequence b) in theorem 1.7 simply by taking the (-1)- eigenspace for complex conjugation in all terms of the exact sequence in theorem 8.19 for $F = \mathbb{C}$. Thus we have a natural identification

$$\mathbb{R} \otimes (\mathbb{R}/\mathbb{Z}) \cong \wedge^2(\mathbb{C}^\times)^-$$

via the map $r \otimes \theta/2\pi \mapsto -e^r \wedge e^{i\theta}$, and it remains to check that the Dehn-invariant is given by $2\lambda^-$. This can be done by comparing the spectral sequences for the Lusztig exact sequences for both $\mathcal{H}^n, \bar{\mathcal{H}}^n$ and $\partial \mathcal{H}^n$. Alternatively first notice that via $\mathcal{P}(\partial \mathcal{H}^3) \cong \mathcal{P}_{\mathbb{C}}^-$ the symbol $\{z\} = [(\infty, 0, 1, z)]$ represents the totally asymptotic 3-simplex with dihedral angles α, β, γ at 3 pairs of "opposite" edges such that $\alpha = $ Arg $z, \beta = $ Arg$(1 - z)$ and $\alpha + \beta + \gamma = \pi$. By an easy calculation $\lambda^-\{z\} \in \mathbb{R} \otimes (\mathbb{R}/\mathbb{Z})$ is then given by

$$D(\infty, 0, 1, z) = 2\lambda^-\{z\} = z \wedge (1 - z) - \bar{z} \wedge (1 - \bar{z})$$

In fact, as observed by W. Thurston (unpublished) the formula (1.2) for the Dehn-invariant makes sense also for a polyhedron with *ideal* vertices if just we delete a horoball around each of these and if for A an edge with such a vertex the length $\ell(A)$ is measured only up to the horosphere. The indeterminacy in this definition vanishes since the sum of angles at edges ending at an ideal vertex is a multiple of π. With this definition one then checks the formula above for $D(\infty, 0, 1, z)$. Again we refer to [**Dupont-Sah, 1982**] for the details.

Remark 2. Similarly to the spherical case the hyperhomology spectral sequence for the Lusztig exact sequence for $\mathcal{H}^n$ (theorem 3.11) has the form

$$E^1_{p,n} = H_*(G(p) \times \mathrm{O}(n-p), \mathrm{St}(\mathcal{H}^p)^t \otimes \mathbb{Z}^t) \Rightarrow H_*(G(n), \mathbb{Z}^t)$$

and in particular we get an edge homomorphism

$$\sigma : H_n(G(n), \mathbb{Z}^t) \to E^\infty_{0,n} \subseteq E^1_{0,n} = \mathcal{P}(\mathcal{H}^n)$$

and again this is described on the chain level by the same formula as for $\bar{\sigma}$ in Remark 3 following theorem 7.14. For $n = 3$ it agrees in $\mathcal{P}^-_{\mathbb{C}} = \mathcal{P}(\mathcal{H}^3)$ with the component σ^- of the map σ in theorem 8.19. However the component σ^+ is not in any obvious way related to "oriented" hyperbolic scissors congruence since by Gerling's theorem (theorem 2.2) $\mathcal{P}(\mathcal{H}^n) = \mathcal{P}(\mathcal{H}^n, \mathrm{SO}^1(1, n))$. As we shall see in the next chapter $\mathcal{P}^+_{\mathbb{C}}$ is rather closely related to $\mathcal{P}(S^3)$. In this connection observe that if a polytope $P \subseteq \mathcal{H}^n$ is the fundamental domain for a discrete co-compact group $\Gamma \subseteq \mathrm{SO}^1(1, n)$ then $M_\Gamma = \Gamma \backslash \mathcal{H}^n$ is an oriented manifold and the fundamental class defines a canonical element $[M_\Gamma] \in H_n(M_\Gamma) \cong H_n(\Gamma)$ and hence by the inclusion $\Gamma \subseteq \mathrm{SO}^1(1, n)$, $[M_\Gamma] \in H_n(\mathrm{SO}^1(1, n))$. It is straight forward to check that $\sigma[M_\Gamma] = [P] \in \mathcal{P}(\mathcal{H}^n)$. That is, for $n = 3$, we have $\sigma^-[M_\Gamma] = [P] \in \mathcal{P}(\mathcal{H}^3)$. But it is not clear how to find $\sigma^+[M_\Gamma]$ in terms of $P \subseteq \mathcal{H}^3$.

Remark 3.   In contrast to corollary 7.7 there is no easy way to relate $\mathcal{P}(\mathcal{H}^{2m-1})$ and $\mathcal{P}(\mathcal{H}^{2m})$. A substitute is theorem 8.5,d) which reduces the calculation of $\mathcal{P}(\mathcal{H}^{2m})$ to $\mathrm{im}(\mathcal{P}(\partial \mathcal{H}^{2m}) \to \mathcal{P}(\bar{\mathcal{H}}^{2m}))$. The first question here is to determine this latter group for $m = 2$.

As a last remark let us note the following consequences of theorems 8.16 and 8.19 (cf. [**Suslin, 1991**], [**Sah, 1989**]):

COROLLARY 8.20.   a) *For F algebraically closed $H_2(\mathrm{Sl}(2, F))$ is a rational vector space and $H_3(\mathrm{Sl}(2, F))$ is the direct sum of a rational vector space and $\mathbb{Q}/\mathbb{Z} \cong H_3(\mu_F)$.*
b) *$\mathcal{P}(\mathcal{H}^3)$ is a uniquely divisible group, i.e. a rational vector space.*

Remark: We shall show that also $\mathcal{P}(S^3)$ is a $\mathbb{Q}$-vector space in the next chapter (corollary 9.19).

CHAPTER 9

Homology of Lie groups made discrete

At this point it must now be clear that there are needed some homology calculations of the isometry groups for the 3 geometries and related Lie groups considered as discrete groups. This subject was first considered in connection with the theory of characteristic classes for foliations (see e.g. [**Kamber-Tondeur, 1968**], [**Cheeger-Simons, 1985**]). It has also been investigated for the case of classical algebraic groups in connection with algebraic K-theory in particular by A. Suslin. We shall skip many proofs, which are rather long. However some of the methods are closely related to the ideas used in the previous chapters. An example of this is the following stability theorem of [**Sah, 1986**]:

THEOREM 9.1. *Consider the unitary group* $G(p,q) = \mathrm{U}(p,q,\mathbb{F}) \subsetneqq \mathrm{Gl}(p+q,\mathbb{F})$, $\mathbb{F} = \mathbb{R}, \mathbb{C}$ *or* $\mathbb{H}$, *of matrices preserving a Hermitian form of signature* (p,q). *The map induced by the inclusion*

$$H_i(G(p,q)) \to H_i(G(p,q+1))$$

is an isomorphism for $i < q$ *and a surjection for* $i = q$.

For the proof let us assume for convenience that $\mathbb{F} = \mathbb{R}$ and $p = 0$, that is, we shall show that

$$H_i(\mathrm{O}(n)) \to H_i(\mathrm{O}(n+1))$$

is an isomorphism for $i < n$ and a surjection for $i = n$. As in chapter 7 let us consider the chain complex $\bar{C}_*(S^n)$ of $\mathrm{O}(n+1)$-modules. The main step in the proof of theorem 9.1 is the following lemma;

LEMMA 9.2. $H_i(\mathrm{O}(n+1)\backslash \bar{C}_*(S^n)) = 0$ *for* $0 < i \le n$.

Proof: As before consider the rank filtration $\bar{C}_*(S^n)^p \subseteq \bar{C}_*(S^n)$ and observe that the natural inclusion $i : S^{n-1} \subseteq S^n$ gives an identification

(9.3) $O(n)\backslash \bar{C}_*(S^{n-1}) = O(n+1)\backslash \bar{C}(S^n)^{n-1}.$

Furthermore the cone construction with the canonical basis vector $e_{n+1} = (0,\ldots,1) \in S^n$ defines a chain homotopy s between i_* and the zero map

$$s(a_0,\ldots,a_n) = (e_{n+1}, a_0,\ldots,a_k), \quad a_0,\ldots,a_n \in S^{n-1}.$$

Hence by (9.3)

$$H_i(O(n+1), \bar{C}_*(S^n)) = 0 \quad \text{for } i \le n-1,$$

and the pair (S^n, S^{n-1}) gives an exact sequence

$$0 \to H_n\left(\frac{\bar{C}^*(S^n)}{O(n+1)}\right) \to H_n\left(\frac{\bar{C}_*(S^n)}{O(n+1)} \bigg/ \frac{\bar{C}_*(S^n)^{n-1}}{O(n+1)}\right) \to$$

$$\to H_{n-1}\left(\frac{\bar{C}_*(S^{n-1})}{O(n)}\right) \to 0$$

However as in the proof of theorem 7.14

$$H_n\left(\frac{\bar{C}_*(S^n)}{O(n+1)} \bigg/ \frac{\bar{C}_*(S^n)^{n-1}}{O(n+1)}\right) =$$
$$H_0(O(n+1)/\operatorname{SO}(n+1), \mathcal{P}(S^n)/\Sigma\mathcal{P}(S^{n-1}))$$

and hence this is annihilated by 2 since the action is the *untwisted* one. However $\mathcal{P}(S^n)$ is 2-divisible, hence the above group vanishes which proves the lemma by the above exact sequence. $\square$

Proof of theorem 9.1.   Since $\bar{C}_*(S^n)$ is acyclic by lemma 3.6 we get a hyperhomology spectral sequence

$$E^1_{p,q} = H_p(O(n+1), \bar{C}_q(S^n)) \Rightarrow H_{p+q}(O(n+1), \mathbb{Z}).$$

Let us prove the theorem by induction and observe that it is true for $n = 1$. In the above spectral sequence we then have by Shapiro's lemma and the induction hypothesis that

$$E^1_{p,q} \cong \frac{\bar{C}_q(S^q)}{O(q+1)} \otimes H_p(O(n)) \quad \text{for } q < n-p, \quad p \le n,$$

and for $q = n - p$ the coefficient group is either $H_p(O(n))$ or $H_p(O(p))$ mapping surjectively to $H_p(O(n))$. Hence by the argument of lemma

9.2

$$E^2_{p,q} \cong \begin{cases} H_p(\mathrm{O}(n)) & q = 0, \\ 0 & 0 < q \leq n - p. \end{cases}$$

(A careful inspection of the coefficients is needed for $q = n - p$ cf. [**Sah, 1986**]). From this it follows that $H_i(\mathrm{O}(n+1)) \cong H_i(\mathrm{O}(n))$ for $i < n$ and $H_n(\mathrm{O}(n+1))$ is a quotient of $H_n(\mathrm{O}(n))$. This establishes the induction and ends the proof of the theorem. $\qquad\square$

For D a division ring with infinite center (e.g. $D = \mathbb{R}, \mathbb{C}$ or $\mathbb{H}$) one can use the same kind of arguments as in the proof of theorem 9.1 but for the action of $\mathrm{Gl}(n+1, D)$ on $D^{n+1} - \{0\}$ and one obtains the following result due to A. Suslin (see [**Sah, 1986**, appendix B] for details of the proof):

THEOREM 9.4. *Let D be a division ring with an infinite center. Then*

a) $H_i(\mathrm{Gl}(n, D)) \to H_i(\mathrm{Gl}(n+1, D))$ *is bijective for $i \leq n$.*
b) *The inclusion $\mathrm{Gl}(1, D) \times \cdots \times \mathrm{Gl}(1, D) \to \mathrm{Gl}(n, D)$ induces a surjection*

$$H_1(\mathrm{Gl}(1, D)) \otimes \cdots \otimes H_1(\mathrm{Gl}(1, D)) \overset{\wedge}{\twoheadrightarrow}$$

$$\frac{H_n(\mathrm{Gl}(n, D))}{\mathrm{im}(H_n(\mathrm{Gl}(n-1), D) \to H_n(\mathrm{Gl}(n, D)))}$$

Remark 1.   For $D = F$ an algebraically closed field $\mathrm{Gl}(n, F) = \mathrm{Sl}(n, F) \times F^\times$. Hence a) is true also for $\mathrm{Gl}(n)$ replaced by $\mathrm{Sl}(n)$ in this case. Furthermore it follows from b) in the same case that for $n \geq 2$ the map $\mathrm{Sl}(n-1) \times \mathrm{Gl}(1) \to \mathrm{Sl}(n)$ sending $(g, \lambda) \mapsto \begin{pmatrix} \lambda^{-1}g & 0 \\ 0 & \lambda \end{pmatrix}$ induces a surjection on H_n. We shall return to this in chapter 13.

Remark 2.   For $D = F$ an infinite field [**Suslin, 1984**] identified the quotient $H_n(\mathrm{Gl}(n, F)) / \mathrm{im}(H_n(\mathrm{Gl}(n-1, F) \to H_n(\mathrm{Gl}(n, F)))$ with a direct summand $K^M_n(F)$, the "Milnor K-group", or the "decomposable part" of the algebraic K-group of D. Quillen $K_n(F) = \pi_n(B\,\mathrm{Gl}(\infty, F)^+), n > 0$. Notice that $K_1(F) \cong F^\times = H_1(\mathrm{Gl}(n, F))$

for all $n \geq 1$. Also for F algebraically closed it follows by the Hurevicz theorem together with theorems 9.4 and 8.19 that for $n \geq 2$:

(9.5)
$$K_2(F) = K_2^M(F) = H_2(\mathrm{Sl}(n, F)) \cong \wedge_{\mathbb{Z}}^2(F^\times)/\{z \wedge (1 - z) \mid z \in F - \{0\}\}$$

Similarly (cf. chapter 15 or [**Sah, 1986**, appendix B]), again for F algebraically closed,

(9.6) $K_3(F) \cong H_3(\mathrm{Sl}(n, F)) = H_3(\mathrm{Sl}(2, F)) \oplus K_3^M(F)$,   for $n \geq 3$,

so that there is a natural identification of $H_3(\mathrm{Sl}(2, F))$ with $K_3(F)^{\mathrm{indec.}}$, the "indecomposable part" of $K_3(F)$. As we shall see these results are particularly important for the study of scissors congruences in spherical and hyperbolic 3-space.

For the Euclidean group $E(n) = T(n) \rtimes O(n)$ the statement analogous to theorem 9.1 is the following

THEOREM 9.7. *The inclusion map* $O(n) \subseteq E(n)$ *induces injective maps*

$$H_i(O(n)) \to H_i(E(n))$$

for all i.

This map is surjective, hence bijective for $i \leq n$.

For the proof we again analyze the spectral sequence for the action of $E(n)$ on the acyclic complex $C_*(E^n)$ of all $(k+1)$-tuples ("k-simplices") $(a_0, \ldots, a_k)$ of points in E^n and we notice that the isotropy subgroup of a point is just $O(n)$. Theorem 9.7 therefore follows in the same way as for the proof of theorem 9.1 just we prove the following:

LEMMA 9.8. $E(n)\backslash C_*(E^n)$ *is* n-*acyclic.*

Proof: For this let us consider the subcomplex $C_*^{\mathrm{gen}}(E^n)$ of "generic" simplices $(a_0, \ldots, a_k)$ with vertices in *general position* , i.e. for every $p \leq n$ there is no subset $(a_{i_0}, \ldots, a_{i_p})$ contained in a $(p-1)$-dimensional affine subspace. Then the lemma clearly follows from the following to properties of the natural map

$$i_* : H_k(E(n)\backslash C_*^{\mathrm{gen}}(E^n)) \to H_k(E(n)\backslash C_*(E^n)) :$$

(9.9,i) i_* is the zero map for $0 < k \leq n$,

(9.9,ii) i_* is a surjection in all dimensions.

To see this recall that the identity: $C_*(E^n) \to C_*(E^n)$ is chain homotopic to the "barycentric subdivision" sd : $C_*(E^n) \to C_*(E^n)$ defined inductively on a simplex $\sigma = (a_0, \ldots, a_k)$ by

(9.10) $\mathrm{sd}(\sigma) = c_\sigma * (\mathrm{sd}(\partial\sigma))$

where c_σ is the barycenter of σ and

$$c_\sigma * (b_0, \ldots, b_l) = (c_\sigma, b_0, \ldots, b_l).$$

Now, if σ is a generic simplex of dimension $\leq n$ then in (9.10) we let c_σ denote the *circumcenter* of σ i.e. the unique point in $\mathrm{span}(\sigma)$ having equal distance from all the vertices $a_0, \ldots, a_k$. With this construction $i : C_k^{\mathrm{gen}}(E^n) \to C_k(E^n)$ is chain homotopic in dimensions $k \leq n$ to sd $: C_k^{\mathrm{gen}}(E^n) \to C_k(E^n)$ where

$$\mathrm{sd}(\sigma) = \sum_{\sigma=\sigma_0 \supset \cdots \supset \sigma_k} \pm(c_{\sigma_0}, \ldots, c_{\sigma_k})$$

However in this sum the terms where $\sigma_{k-1} = (a_i, a_j), \sigma_k = (a_i)$ cancel with the terms where $\sigma_{k-1} = (a_i, a_j), \sigma_k = (a_j)$ since they are congruent by the reflection in the hyperplane bisecting (a_i, a_j) and since they occur with opposite signs. This proves (9.9,i). For (9.9,ii) we have a similar construction of a subdivision operator sd : $C_*(E^n) \to C_*^{\mathrm{gen}}(E^n)$ by the formula (9.10) but this time we choose as c_σ some point in general position to previous choices of c_τ for $\tau \subset \sigma$.   □

As an application we can now prove the vanishing result needed in the proof of Sydler's theorem (theorem 6.1):

THEOREM 9.11. $H_p(O(n), \wedge_{\mathbb{Q}}^q(\mathbb{R}^n)) = 0$ *if* $p + q \leq n, \quad q > 0$.

Proof: The Hochschild-Serre spectral sequence for the extension

$$0 \to T(n) \to E(n) \to O(n) \to 1$$

has

$$E_{p,q}^2 = H_p(O(n), H_q(\mathbb{R}^n)) \Rightarrow H_{p+q}(E(n)),$$

and by proposition 4.7, ii), we have $H_q(\mathbb{R}^n) \cong \wedge_{\mathbb{Q}}^q(\mathbb{R}^n)$ for $q \geq 1$. As in chapter 4 the dilatation $\mu_a : \mathbb{R}^n \to \mathbb{R}^n$ by $a \in \mathbb{N}$ gives an endomorphism commuting with the differentials $d^r : E_{p,q}^r \to E_{p-r,q+r-1}^r$. Again μ_a

induces multiplication by a^q on $\wedge_{\mathbb{Q}}^q(\mathbb{R}^n)$ so that $d^r = 0$ for $r \geq 2$. Hence for $p + q = k \leq n$ we conclude from theorem 9.7:

$$H_k(O(n)) = H_k(E(n)) \cong H_k(O(n)) \oplus \bigoplus_{q=1}^{k} H_{n-q}(O(n), \wedge_{\mathbb{Q}}^q(\mathbb{R}^n))$$

from which the theorem clearly follows. $\square$

For the group $O^1(1, n)$ of isometries of hyperbolic space we have the following result of [**Bökstedt-Brun-Dupont, 1998**] similar to theorem 9.7:

THEOREM 9.12. *The inclusion map* $O(n) \subseteq O^1(1, n)$ *induces an isomorphism* $H_i(O(n)) \to H_i(O^1(1, n))$ *for* $i < n$ *and a surjection for* $i = n$.

The proof of this theorem follows the same strategy as the proof of theorem 9.7 and similarly we must prove

LEMMA 9.13. $O^1(1, n) \backslash C_*(\mathcal{H}^n)$ *is* n-*acyclic.*

However for the proof of this we encounter the difficulty that a simplex $(a_0, \ldots, a_k)$ in $\mathcal{H}^k$ with the vertices in general position need not in general have a *circumcenter* as seen in the figure below of a triangle in $\mathcal{H}^2$ where two of the bisecting normals do not intersect.

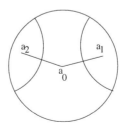

On the other hand if we subdivide the simplices the geometry gets closer to the Euclidean one which makes the existence of circumcenters likely for the smaller simplices. However the usual barycentric subdivision does not work since it creates "long and thin" simplices. The solution is to use the so-called "edgewise subdivision" of the standard affine simplex Δ^k and use the usual parametrisation of a hyperbolic simplex $\sigma = (a_0, \ldots, a_k)$ as in chapter 2:

$$f_\sigma(t) = \frac{a(t)}{\sqrt{-\langle a(t), a(t) \rangle_-}}, \quad a(t) = a(t_0, \ldots, t_k) = \sum_i t_i a_i,$$

$$t \in \Delta^k \subseteq \mathbb{R}^{k+1},$$

to transport the subdivision to $\mathcal{H}^n \subset \mathbb{R}^{n+1}$. Now lemma 9.13 follows similarly to the proof of lemma 9.8 from:

LEMMA 9.14. *There is a* $O^1(1, n)$-*equivariant chain map* $\mathrm{Sd}_{\mathcal{H}^n}$: $C_*(\mathcal{H}^n) \to C_*(\mathcal{H}^n)$ *chain homotopic to the identity such that for any generic k-simplex* $\sigma, k \leq n$, *there is an integer* M *such that every simplex in* $(\mathrm{Sd}_{\mathcal{H}^n})^m(\sigma)$, *for* $m \geq M$, *possesses a (unique) circumcenter in* span σ.

We refer to [**Bökstedt-Brun-Dupont, 1998**, theorem 3.3.1] for the proof of this lemma. Let us just give the description of "edgewise subdivision" $\mathrm{Sd} : C_*(E^n) \to C_*(E^n)$. Then for any affine simplex $\tau = (x_0, \ldots, x_k)$ in $E^n, k \leq n$, it turns out that each simplex in $\mathrm{Sd}(\tau)$ is contained in $|\tau|$ so that we can define for σ a *hyperbolic simplex* as above

$$\mathrm{Sd}_{\mathcal{H}^n}(\sigma) = (f_\sigma')_*(\mathrm{Sd}(\Delta_0^k))$$

where we have used the following notation: For $\{e_i\}$ the standard basis of $\mathbb{R}^n$

$$\Delta_0^k = (0, e_1, e_1 + e_2, \ldots, e_1 + \cdots + e_k) = (d_0, \ldots, d_k)$$

is the "standard" k-simplex in E^m. Hence with f_σ as defined above we put

$$f_\sigma'\left(\sum_{i=0}^k t_i d_i\right) = f_\sigma(t_0, \ldots t_k) \quad t = (t_0, \ldots, t_k) \in \Delta^k,$$

and

$$(f_\sigma')_*(x_0, \ldots, x_k) = (f_\sigma'(x_0), \ldots, f_\sigma'(x_k)), \quad x_i \in |\Delta_0^k|.$$

For the construction of Sd we identify as in chapter 4, $C_*(E^n)$ with the bar complex $B_*(V), V = \mathbb{R}^n$, via the identification

$$(x_0, \ldots, x_k) \in C_*(E^n) \leftrightarrow u_0[u_1| \ldots |u_k] \in B_*(V)$$

where

$$x_0 = u_0, \quad x_1 = u_0 + u_1, \ldots, \quad x_k = u_0 + u_1 + \cdots + u_k.$$

In particular the standard k-simplex above is just

$$\Delta_0^k = 0[e_1|e_2|\ldots|e_k],$$

that is as a set

$$\Delta_0^k = \{(s_1, s_2, \ldots, s_k, 0, 0, \ldots, 0) \mid 1 \geq s_1 \geq s_2 \geq \cdots \geq s_k \geq 0\}.$$

As usual, we shall identify a simplex $v_0[v_1|v_2|\ldots|v_k]$ with the affine map $\Delta_0^k \to V$ sending 0 to v_0 and e_i to v_i, $i = 1, 2, \ldots, k$.

DEFINITION 9.15. The *edgewise subdivision* is the chain map Sd : $B_*(V) \to B_*(V)$ given by the composite

$$\text{Sd} : B_*(V) \overset{(\frac{1}{2})_*}{\to} B_*(V) \overset{AW}{\to} B_*(V) \overset{EZ}{\to} B_*(V \times V) \overset{(+)_*}{\to} B_*(V)$$

with the following notation: $(\frac{1}{2})_*$ and $(+)_*$ are induced by the homomorphism

$$V \overset{\frac{1}{2}}{\to} V \qquad \text{respectively} \qquad V \times V \overset{+}{\to} V$$
$$v \mapsto \frac{1}{2}v \qquad\qquad\qquad\qquad (v, w) \mapsto v + w,$$

AW is the Alexander-Whitney map (cf. [**MacLane, 1963**, chap. VIII §9 exercise 1]

$$AW(v_0[v_1|\ldots|]) = \sum_{p=0}^{k} v_0[v_1|\ldots|v_p] \otimes (v_0 + \cdots + v_p)[v_{p+1}|\ldots|v_k]$$

and EZ is the Eilenberg-Zilber shuffle map (cf. chapter 4):

$$EZ(v_0[v_1|\ldots|v_p] \otimes w_0[w_1|\ldots|w_q]) = \sum_{\sigma} \text{sign } \sigma \ (v_0, w_0)[u_{\sigma 1}|\ldots|u_{\sigma k}]$$

where σ runs through all (p, q)-shuffles (cf. MacLane up cit.) and where $u_i = (v_i, 0)$ for $i = 1, \ldots, p$ and $u_i = (0, w_i)$ for $i = p + 1, \ldots, p + q = k$.

Remark: As noted in chapter 4, $EZ(v_0[v_1|\ldots|v_p] \otimes w_0[w_1|\ldots|w_q])$ is represented geometrically by a triangulation of the (p, q)-prism

$$v_0[v_1|\ldots|v_p] \times w_0[w_1|\ldots|w_q] \subseteq V \times V.$$

Hence, in the notation of that chapter

$$v_0[v_1|\ldots|v_p]\wedge w_0[w_1|\ldots|w_q] =_{\text{def}} (+)_* EZ(v_0[v_1|\ldots|v_p]\otimes w_0[w_1|\ldots|w_q])$$

corresponds to a triangulation of the polytope

$$v_0[v_1|\ldots|v_p] + w_0[w_1|\ldots|w_q] \subset V$$

(the "Minkowski sum"). In particular

$$\wedge \circ AW(v_0[v_1|\ldots|v_k]) = \sum_{p=0}^{k} v_0(v_1|\ldots|v_p] \wedge (v_0 + \cdots + v_p)[v_{p+1}|\ldots|v_k]$$

corresponds to (a triangulation of) a decomposition of the simplex $2v_0[2v_1|\ldots|2v_k]$ into prisms (cf. proof of proposition 4.12 and the figure below). It follows that

$$\text{Sd}(v_0[v_1|\ldots|v_k]) =$$

(9.16)
$$\sum_{p=0}^{k} \frac{1}{2}v_0 \left[\frac{1}{2}v_1|\ldots|\frac{1}{2}v_k\right] \wedge \frac{1}{2}(v_0 + \cdots + v_p) \left[\frac{1}{2}v_{p+1}|\ldots|\frac{1}{2}v_k\right]$$

is a triangulation of $v_0[v_1|\ldots|v_k]$.

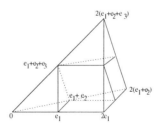

$$\wedge \circ AW(\Delta_0^3)$$

If we now identify the chain complex of affine simplices $C_*(\Delta_0^n)$ of Δ_0^n with the corresponding subcomplex of $B_*(V)$, then we obtain the following lemma, which follows easily from the above discussion (cf. proof of proposition 4.12).

LEMMA 9.17.   i) Sd *defines a chain map*

$$\text{Sd} : C_*(\Delta_0^n) \to C_*(\Delta_0^n)$$

ii) Sd *is natural with respect to affine maps*

iii) *There is a natural chain homotopy of* Sd *to the identity, again natural with respect to affine maps.*

iv) *All simplices in* $\mathrm{Sd}(\Delta_0^n)$ *are isometric to* $\frac{1}{2}\Delta_0^n$ *by a permutation of the basis vectors* $\{e_i\}$ *followed by a translation by a vector of the form* $\frac{1}{2}(e_{i_1} + \cdots + e_{i_l})$.

The last statement in this lemma makes it possible to control in a uniform way the "shape" of the simplices occurring in the corresponding subdivision $\mathrm{Sd}_{\mathcal{H}^n}(\sigma)$ of a hyperbolic simplex. For details we refer to [**Bökstedt-Brun-Dupont, 1998**] (where however Sd is defined in a different but equivalent way). This ends our outline of the proof of theorem 9.12. For our applications to scissors congruences let us state the following.

COROLLARY 9.18.   a) *The inclusion* $\mathrm{O}(3) \subseteq \mathrm{O}^1(1,3)$ *induces an isomorphism* $H_k(\mathrm{O}(3)) \overset{\cong}{\to} H_k(\mathrm{O}^1(1,3))$ *for* $k \leq 3$.
   b) *The inclusion* $\mathrm{SU}(2) \subseteq \mathrm{Sl}(2,\mathbb{C})$ *induces an isomorphism*

$$H_k(\mathrm{SU}(2)) \overset{\cong}{\to} H_k(\mathrm{Sl}(2,\mathbb{C}))^+, \quad k \leq 3,$$

where + *indicates the invariant part for the involution induced by complex conjugation.*

Proof: a) By theorem 9.12 the only problem is the injectivity for $k = 3$. But by the diagram induced by inclusion maps

$$
\begin{array}{ccc}
H_3(\mathrm{O}(3)) & \longrightarrow & H_3(\mathrm{O}(1,3)) \\
\downarrow & & \downarrow \\
H_3(\mathrm{O}(4)) & \longrightarrow & H_3(\mathrm{O}(1,4))
\end{array}
$$

it suffices to show that $i : \mathrm{SO}(3) \to \mathrm{SO}(4)$ induces an injection. But writing $\mathrm{SO}(4) = (S_1 \times S_2)/\{\pm 1\}$ as in chapter 7, we get a splitting of i by the projection onto the first factor.

b) By the Hochschild-Serre spectral sequence we now get an isomorphism

$$H_k(\mathrm{Pin}(3)) \to H_k(\mathrm{Pin}(1,3)) \quad \text{for} \quad k \leq 3.$$

Now the result follows from the isomorphisms $\text{Pin}(3) \cong \{\pm 1\} \times \text{SU}(2)$ and $\text{Pin}(1,3) \cong \text{Sl}(2,\mathbb{C}) \rtimes \{1,\tau\}$ where τ is the involution on $\text{Sl}(2,\mathbb{C})$ given by $\tau(g) = (\bar{g}^t)^{-1}$, together with corollary 8.20. Notice that the involution τ is conjugate (using the matrix $\begin{pmatrix} 0 & 1 \\ -1 & 0 \end{pmatrix}$) to the involution $g \mapsto \bar{g}$ so in homology these involutions have the same invariant parts. $\qquad\square$

COROLLARY 9.19.   a) $H_2(\text{SU}(2))$ *is a rational vector space and* $H_3(\text{SU}(2))$ *is a direct sum of a rational vector space and* $\mathbb{Q}/\mathbb{Z} \cong H_3(\mu_\mathbb{C})$.
   b) $\mathcal{P}(S^3)$ *is a rational vector space.*

Proof: a) clearly follows from corollary 8.20. b) now follows from the exact sequence theorem 1.7, c) together with a) and the fact that $\mathbb{Q}/\mathbb{Z} \subseteq H_3(\text{SU}(2))$ is mapped by σ to the group of rational lunes in $\mathcal{P}(S^3)/\mathbb{Z}$ (cf. remark 2 following theorem 7.14). $\qquad\square$

In view of corollary 9.18 and the similarity of the sequences in theorems 1.7, c) and 8.19 it is natural to expect some relation between $\mathcal{P}(S^3)$ and $\mathcal{P}_\mathbb{C}^+$, the invariant part of $\mathcal{P}_\mathbb{C}$ by complex conjugation. We shall briefly describe this relation referring to [**Dupont-Parry-Sah, 1988**, §5] for details. First observe that the Hopf map of $S^3 \subseteq \mathbb{C}^2$ to the Riemann sphere $P^1(\mathbb{C}) = S^2$ is $\text{U}(2)$-equivariant. By comparing the spectral sequences for $\text{U}(2)$ and $\text{SO}(4)$ we get the horizontal map in the following diagram to be an isomorphism,

$$H_3(\text{U}(2)\backslash \bar{C}_*(S^3)) \xrightarrow{\;\cong\;} H_3(\text{SO}(4)\backslash \bar{C}_*(S^3)) \cong \mathcal{P}(S^3)/\Sigma\mathcal{P}(S^2)$$

$$\Big\downarrow{\scriptstyle \text{Hopf map}} \nearrow \xi$$

$$H_3(\text{Sl}(2,\mathbb{C})\backslash \bar{C}_*(S^2)) = \mathcal{P}_\mathbb{C}$$

thus defining the map ξ. Note that we have used theorem 7.14. Furthermore consider the subgroup $\mathcal{P}(S^1) * \mathcal{P}(S^1) \subseteq \mathcal{P}(S^3)$ generated by all orthogonal joins of arcs on S^1, i.e. simplices of the form (a_0, a_1, a_2, a_3) with span$\{a_0, a_1\}$ orthogonal to span$\{a_2, a_3\}$. Then using corollary 9.18 one proves the following (cf. [**Dupont-Parry-Sah, 1988**, theorem 5.15]:

THEOREM 9.20. *There is an exact sequence*

$$0 \to \mathcal{P}(S^1) * \mathcal{P}(S^1) \to \mathcal{P}(S^3) \xrightarrow{\xi} \mathcal{P}_{\mathbb{C}}^+ \xrightarrow{\lambda_{\mathbb{R}}} \Lambda_{\mathbb{Z}}^2(\mathbb{R}^{\times}) \to 0$$

where $\lambda_{\mathbb{R}}(\{z\}) = |z| \wedge |1 - z|$.

We shall return to this relation between $\mathcal{P}(S^3)$ and $\mathcal{P}_{\mathbb{C}}^+$ in chapter 11.

As in corollary 9.19 we are interested in the structure of the homology groups, e.g. if they have torsion or are divisible. In this context we should mention the following conjecture of E. Friedlander and J. Milnor (cf. [**Milnor, 1983**]; see also [**Roger, 1975**]): Recall that for an arbitrary group G the homology $H_*(G)$ which we have considered above is in a natural way identified with the singular homology of its classifying space. For G a Lie group let us denote the classifying space of the underlying discrete group by BG^δ to distinguish it from the classifying space BG in the topological sense. Thus $H_*(BG^\delta, \mathbb{Z}) = H_*(G, \mathbb{Z})$ in the above notation. We can now state:

The Friedlander-Milnor Conjecture (FMC). The canonical map $BG^\delta \to BG$ induces an isomorphism of homology with mod p coefficients $p > 0$.

The point of this conjecture is that the homology of BG is rather well understood for most Lie groups in particular for the classical ones and hence makes it possible to calculate $H_*(G, \mathbb{Z}/p), p > 0$. By the Hochschild-Serre spectral sequence it is easy to see that the FMC is true for a Lie group G if it is true for its connected component G_0. Similarly FMC is true for a connected Lie group G if and only if it is true for its universal covering group.

EXAMPLE 9.21. $G = \mathbb{R}$. By proposition 4.7, $H_*(\mathbb{R}) = \Lambda_{\mathbb{Q}}^*(\mathbb{R})$ is a rational vector space hence by the universal coefficient theorem $H_*(\mathbb{R}, \mathbb{Z}/p) = 0$. On the other hand the classifying space in the continuous topology $B\mathbb{R}$ is contractible since $\mathbb{R}$ is contractible; hence also $H_*(B\mathbb{R}, \mathbb{Z}/p) = 0$ and we have thus proved FMC in this case.

By induction it is now straightforward to establish FMC for all solvable Lie groups and in fact reduce the problem to semi-simple groups. For the classical groups let us quote without proof the following deep result of [**Suslin, 1986**], [**Suslin, 1991**]:

THEOREM 9.22. FMC *is true for* $\mathrm{Gl}(n, F)$ *and* $\mathrm{Sl}(n, F)$, $F = \mathbb{R}, \mathbb{C}$, *in the stable range given in theorem 9.4, i.e. for* H_i *with* $i \leq n$.

COROLLARY 9.23. $H_3(\mathrm{Sl}(n, \mathbb{C})) \cong H_3(\mu_{\mathbb{C}}) \oplus \mathbb{Q}$-*vector space for* $n \geq 2$. *Here* $H_3(\mu_{\mathbb{C}}) \cong \mathbb{Q}/\mathbb{Z}$.

Proof: We have noted the last statement before in theorem 8.19. For the first statement observe that by (9.6) we can take $n \geq 3$ and we observe that $B \mathrm{Sl}(n, \mathbb{C})$ for $n \geq 2$ is 3-connected with $H_4(B \mathrm{Sl}(n, \mathbb{C})) \cong \mathbb{Z}$. In fact the generator is dual to the second Chern class c_2 for complex vector bundles. Hence by theorem 9.22, also $H_i(\mathrm{Sl}(n, \mathbb{C}), \mathbb{Z}/p) = 0$ for $i \leq 3$ and $H_4(\mathrm{Sl}(n, \mathbb{C}), \mathbb{Z}/p) \cong \mathbb{Z}/p$. But since $H_3(\mu_{\mathbb{C}}) \subseteq H_3(\mathrm{Sl}(n, \mathbb{C}))$ already gives rise to a copy of $\mathbb{Z}/p$ in H_4 there can be no other torsion factors in $H_3(\mathrm{Sl}(n, \mathbb{C}))$. Since $H_3(\mathrm{Sl}(n, \mathbb{C}))$ is p-divisible by the exact sequence for the coefficient sequence $\mathbb{Z} \xrightarrow{p.} \mathbb{Z} \to \mathbb{Z}/p$, the result now follows. $\square$

Remark 1   [**Suslin, 1983**] has also proved that if $F \subseteq \mathbb{C}$ is an algebraically closed subfield then the inclusion $\mathrm{Sl}(n, F) \to \mathrm{Sl}(n, \mathbb{C})$ induces an isomorphism of homology groups with finite coefficients in the stable range. Hence corollary 9.23 is valid also for $\mathbb{C}$ replaced by F.

Remark 2.   FMC is also known in the stable range for other families of classical groups e.g. the orthogonal groups $O(n, n)$ (see [**Sah, 1989**]). However except in dimensions ≤ 2 (where it is verified for almost all simple groups, c.f. [**Sah, 1989**]) little is known for *compact* simple groups. Corollary 9.19 shows FMC for $SU(2)$ in dimensions ≤ 3 and hence by theorem 9.1 in the same dimensions for the families $\mathrm{Spin}(n), SU(n)$ and $Sp(n)$ (if we use the coincidences of groups of low dimensions in these families). Finally note that by theorem 9.12 FMC is true for $O(n)$ in the stable range $i < n$ if and only if it is true for $O^1(1, n)$ in the same range and one could hope gradually to reduce it in the same manner to the known case of $O(n, n)$.

Remark 3. For the purpose of calculating scissors congruence groups FMC gives however, even if true, no information about possible $\mathbb{Q}$-vector spaces in $H_*(G)$ and these can be quite large as seen already in example 9.21.

CHAPTER 10

Invariants

Historically the notion of scissors congruences occurred in connection with the concept of "volume", but in the above homological treatment of the hyperbolic and spherical case the volume occurred very rarely. The reason is that basically the problem of scissors congruence is an algebraic one whereas "volume" in the non-Euclidean case is given by transcendental functions. This was observed by B. Jessen (cf. [**Jessen, 1973**],[**Jessen, 1978**]) who stated the non-Euclidean Hilbert's 3rd Problem (see chapter 1) but nevertheless found a positive answer unlikely. We shall return to this in the next chapter. In this chapter we shall relate "volume" in 3-dimensional non-Euclidean geometry to the invariants for flat $\mathrm{Sl}(2,\mathbb{C})$-bundles introduced in differential geometry, the *Cheeger-Chern-Simons classes* (cf. [**Chern-Simons, 1974**], [**Cheeger-Simons, 1985**]). We shall also give formulas for these in terms of the *dilogarithmic function* . Also we shall see that we get some further invariants for s.c. if we twist the Cheeger-Chern-Simons invariants with a field automorphism of $\mathbb{C}$.

First let us summarise our results so far about 3-dimensional non-Euclidean scissors congruences in the following 2 exact sequences (theorem 1.7 and corollary 9.18)

$(10.1,-)$

$$0 \to H_3(\mathrm{Sl}(2,\mathbb{C}))^- \xrightarrow{\sigma} \mathcal{P}(\mathcal{H}^3) \xrightarrow{D} \mathbb{R}\otimes\mathbb{R}/\mathbb{Z} \to H_2(\mathrm{Sl}(2,\mathbb{C}))^- \to 0$$

$(10.1,+)$

$$0 \to H_3(\mathrm{Sl}(2,\mathbb{C}))^+ \xrightarrow{\sigma} \mathcal{P}(S^3)/\mathbb{Z} \xrightarrow{D} \mathbb{R}\otimes\mathbb{R}/\mathbb{Z} \to H_2(\mathrm{Sl}(2,\mathbb{C}))^+ \to 0$$

Also let us recall from corollary 9.23 that the only torsion occurring in the homology groups in $(10.1,\pm)$ is a copy of $\mathbb{Q}/\mathbb{Z} = H_3(\mu_{\mathbb{C}}) \subseteq$

$H_3(\mathrm{Sl}(2,\mathbb{C}))^+$, and everything else are rational vector spaces. We shall indicate a proof of the following:

THEOREM 10.2. *The second Cheeger-Chern-Simons class $\hat{C}_2$ for flat* $\mathrm{Sl}(2,\mathbb{C})$ *bundles is the homomorphism* $\hat{C}_2 : H_3(\mathrm{Sl}(2,\mathbb{C})) \to \mathbb{C}/\mathbb{Z}$ *given by*

$$\langle \hat{C}_2, z \rangle = \frac{1}{2\pi^2} \left(\mathrm{Vol}_{S^3} \circ \sigma(z_+) + \frac{i}{2} \mathrm{Vol}_{\mathcal{H}_3} \circ \sigma(z_-) \right).$$

Remark: Note the $1/2$ in the hyperbolic case. We shall explain this below.

First let us discuss the Cheeger-Chern-Simons classes more generally for flat $\mathrm{Gl}(n,\mathbb{C})$-bundles. We shall restrict to this case although much of the discussion is valid for other Lie groups. Let $E \to M$ be a principal $\mathrm{Gl}(n,\mathbb{C})$-bundle with a *flat* connection A. Then by classical Chern-Weil theory (see e.g. [**Dupont, 1978**]) the k-th Chern class $c_k(E) \in H^{2k}(M,\mathbb{Z})$ goes to zero in $H^{2k}(M,\mathbb{C})$ since it is represented by the k-th Chern polynomial C_k in the curvature form of A. Hence by the Bockstein exact sequence

$$\cdots \to H^{2k-1}(M,\mathbb{C}/\mathbb{Z}) \overset{\beta}{\to} H^{2k}(M,\mathbb{Z}) \to H^{2k}(M,\mathbb{C}) \to \cdots$$

it follows that $c_k(E)$ is in the image of the Bockstein homomorphism β. The k-th *Cheeger-Chern-Simons class* which were originally defined in [**Chern-Simons, 1974**] or [**Cheeger-Simons, 1985**], gives a canonical choice

$$\hat{C}_k(A) \in H^{2k-1}(M,\mathbb{C}/\mathbb{Z})$$

such that $-\beta(\hat{C}_k(A)) = c_k(E)$ (cf. chapter 12 below). Since a flat bundle is determined by the holonomy representation of the fundamental group into $\mathrm{Gl}(n,\mathbb{C})^\delta$, it follows that the classifying space for such bundles is $B\,\mathrm{Gl}(n,\mathbb{C})^\delta$. In this way we obtain a universal class

$$\hat{C}_k \in H^{2k-1}(B\,\mathrm{Gl}(n,\mathbb{C})^\delta, \mathbb{C}/\mathbb{Z}) \cong H^{2k-1}(\mathrm{Gl}(n,\mathbb{C}), \mathbb{C}/\mathbb{Z})$$
$$\cong \mathrm{Hom}(H_{2k-1}(\mathrm{Gl}(n,\mathbb{C})), \mathbb{C}/\mathbb{Z}),$$

Here for G any group and M a trivial G-module

$$H^*(G,M) = H(\mathrm{Hom}_{\mathbb{Z}[G]}(B_*(G), M))$$
$$= H(\mathrm{Hom}_{\mathbb{Z}}(\bar{B}_*(G), M))$$

with $(B_*(G), \partial_G)$ the bar resolution as in chapter 4.

Now let G be a Lie group (e.g. $G = \mathrm{Gl}(n, \mathbb{C})$) acting on a manifold V (on the left) and suppose that V is $(q-1)$-connected for some integer q. Then the chain complex $C_*^{\mathrm{sing}}(V)$ of smooth singular simplices in V is in a natural way a chain complex of (left) G-modules, and since it is $(q-1)$-acyclic we can find a chain transformation σ of G-modules up to level q:

$$
\begin{array}{ccccccccc}
\mathbb{Z} & \xleftarrow{\;\varepsilon\;} & B_0(G) & \xleftarrow{\;\partial_G\;} & B_1(G) & \xleftarrow{\;\partial_G\;} & \cdots & \xleftarrow{\;\partial_G\;} & B_q(G) \\
\downarrow & & \downarrow{\scriptstyle\sigma} & & \downarrow{\scriptstyle\sigma} & & & & \downarrow{\scriptstyle\sigma} \\
\mathbb{Z} & \xleftarrow{\quad} & C_0^{\mathrm{sing}}(V) & \xleftarrow{\;\partial\;} & C_1^{\mathrm{sing}}(V) & \xleftarrow{\;\partial\;} & \cdots & \xleftarrow{\;\partial\;} & C_q^{\mathrm{sing}}(V)
\end{array}
$$

(cf. [**MacLane, 1963**, chap. III thm. 6.1]). Furthermore, any two such chain transformations are chain homotopic up to level $q-1$. Given σ as above we define for each G-invariant complex value p-form ω on V with $p \le q$ a cochain $\eth(\omega) \in \mathrm{Hom}(\bar{B}_p(G), \mathbb{C})$ by the integral

$$(10.3) \qquad \eth(\omega)([g_1|\ldots|g_p]) = \int_{\sigma[g_1|\ldots|g_p]} \omega$$

The following proposition is a straightforward application of Stokes' theorem:

PROPOSITION 10.4. *Suppose ω is closed and for $p = q$ that it has integral periods, i.e. $\int_z \omega \in \mathbb{Z}$ for $z \in C_*^{\mathrm{sing}}(V)$ a cycle. Then*

i) *The cochain $\eth(\omega)$ is a cocycle for $p < q$ and a cocycle mod $\mathbb{Z}$ for $p = q$. Furthermore it is a coboundary if $\omega = d\omega'$ for ω' another G-invariant form.*

ii) *The cohomology class of $\eth(\omega)$ in $H^q(G, \mathbb{C})$ for $p < q$, respectively $H^q(G, \mathbb{C}/\mathbb{Z})$ for $p = q$, is independent of choice of σ.*

iii) *If furthermore $H^q(V, \mathbb{Z}) \cong \mathbb{Z}$ and ω respects the generator, then*

$$-\beta([\eth(\omega)]) \in H^{q+1}(G, \mathbb{Z}) = H^{q+1}(BG^\delta, \mathbb{Z})$$

is the obstruction $(q+1)$-cocycle to the existence of a lift $\tilde{\psi}$ in the diagram

$$EG \times_G V$$

where $EG \to BG$ is the universal G-bundle.

iv) *(Naturality) Suppose $\varphi : G' \to G$ is a Lie group homomorphism and $\Phi : V' \to V$ a φ-equivariant differentiable map (i.e. $\Phi(g'v') = \varphi(g')\Phi(v')$ for $g' \in G'$, $v' \in V'$); then*

$$\mathcal{J}(\Phi^*\omega) = \varphi^*\mathcal{J}(\omega).$$

$\square$

Now return to $G = \mathrm{Gl}(n, \mathbb{C})$ and let us take

$$V = \mathrm{Gl}(n, \mathbb{C})/\mathrm{Gl}(k-1, \mathbb{C})$$

which is $(2k-2)$-connected and has $H^{2k-1}(V, \mathbb{Z}) \cong \mathbb{Z}$. Here V is the complexification of the compact homogeneous space $\mathrm{U}(n)/\mathrm{U}(k-1)$ and hence the generator of the $(2k-1)$th cohomology group can be represented in the de Rham cohomology by a *complex* valued G-invariant form ω_k. Thus proposition 10.4 yields a cochain $\mathcal{J}(\omega_k)$ with values in $\mathbb{C}/\mathbb{Z}$ and satisfying

(10.5) $$-\beta([\mathcal{J}(\omega_k)]) = c_k$$

where $c_k \in H^{2j}(B\,\mathrm{Gl}(n, \mathbb{C})^\delta, \mathbb{Z})$ is the pull-back of the universal kth Chern class. In view of (10.5) it is not surprising that in fact

(10.6) $$[\mathcal{J}(\omega_k)] = \hat{C}_k \in H^{2k-1}(B\,\mathrm{Gl}(n, \mathbb{C})^\delta, \mathbb{C}/\mathbb{Z}).$$

For a proof of this, which uses the de Rham theory of differential forms on simplicial manifolds, we refer to [**Dupont-Kamber, 1990**, thm. 5.3].

Remark: The formula (10.3) for $\mathcal{J}(\omega)$ involves a choice of the transformation σ. In the case of $V = G/K$ a symmetric space of non-compact type, i.e. G semi- simple with maximal compact subgroup K, a canonical choice for σ (with $q = \infty$) is provided by letting $\sigma[g_1|\ldots|g_p]$ be the so-called *geodesic simplex* $\Delta(g_1, \ldots, g_p)$ defined inductively as the *geodesic cone* on $g_1\Delta(g_2, \ldots, g_p)$ with cone point $o = \{K\}$ (see e.g.

[**Dupont, 1976**]). Note that a different choice of cone point will give a different but cohomologous cocycle. We shall use this construction in the case of $G = \mathrm{Sl}(2, \mathbb{C})$ acting on $\mathcal{H}^3$.

Proof of theorem 10.2.   As for the real part $\mathrm{Re}\,\hat{C}_2$ it suffices to check the value on

$$H_3(\mathrm{SU}(2)) = H_3(\mathrm{Sl}(2, \mathbb{C}))^+$$

In this case the formula

$$\langle \mathrm{Re}\,\hat{C}_2, z \rangle = \frac{1}{2\pi^2}\,\mathrm{Vol}_{S^3}(\sigma(z))$$

clearly follows from the definition in proposition 10.4 since for $G = \mathrm{SU}(2)$ acting on S^3 in the usual way

$$(10.7) \qquad \mathrm{Re}\,\hat{C}_2 = \mathscr{J}(\omega_2\,|_{S^3}) \text{ and } \omega_2\,|_{S^3} = v_{S^3}/\,\mathrm{Vol}(S^3) = v_{S^3}/2\pi^2,$$

where v_{S^3} denotes the volume form on S^3 (cf. Remark 3 at the end of chapter 7). To determine the imaginary part we take as model for the hyperbolic space $\mathcal{H}^3$ the upper half space in the 3-dimensional subspace of the quaternions $\boldsymbol{H}$ spanned by $1, i, j$:

$$\mathcal{H}^3 = \{x = x_1 + x_2 i + x_3 j \mid x_1, x_2, x_3 \in \mathbb{R}, \quad x_3 > 0\},$$

and $\mathrm{Sl}(2, \mathbb{C})$ acts on $\mathcal{H}^3$ by the usual formula

$$g(x) = (ax + b)(cx + d)^{-1} \quad \text{for} \quad g = \begin{pmatrix} a & b \\ c & d \end{pmatrix}$$

where $\mathbb{C} \subset \mathcal{H}$ is spanned by 1 and i. In particular the map $q : \mathrm{Sl}(2, \mathbb{C}) \to \mathcal{H}^3$ given by

$$q(g) = g(j) = (aj + b)(cj + d)^{-1}$$

gives an equivariant map (with respect to the left action) inducing a diffeomorphism $\mathrm{Sl}(2, \mathbb{C})/\,\mathrm{SU}(2) \approx \mathcal{H}^3$. Now the formula for the imaginary part

$$\langle \mathrm{Im}\,\hat{C}_2, z \rangle = \frac{1}{4\pi^2}(\mathrm{Vol}_{\mathcal{H}^3}(\sigma(z)))$$

clearly follows from the following lemma which is checked by direct calculation (cf. [**Dupont, 1987**, lemma 3.4]).

LEMMA 10.8. *Let $v_{\mathcal{H}^3}$ be the volume form on $\mathcal{H}^3$ then $\mathrm{Im}(\omega_2)$ and $(1/4\pi^2)q^* v_{\mathcal{H}^3}$ are cohomologous in the complex of real left invariant forms on $\mathrm{Sl}(2, \mathbb{C})$.*

□

In chapter 8 we made the identification

$$\mathcal{P}(\mathcal{H}^3) \cong \mathcal{P}(\partial \mathcal{H}^3) \cong \mathcal{P}_\mathbb{C}^-$$

and in these terms $\sigma : H_3(\mathrm{Sl}(2, \mathbb{C})) \to \mathcal{P}_\mathbb{C}^-$ was induced by

$$\sigma([g_1|g_2|g_3]) = \{z\}$$

where z is the cross ratio for the simplex $(\infty, g_1\infty, g_1 g_2\infty, g_1 g_2 g_3\infty)$ in $\partial\mathcal{H}^3$. By taking $o = \infty$ in the construction of geodesic simplices (cf. remark following proposition 10.4) we see from lemma 10.8 that $\mathrm{Im}\,\hat{C}_2$ is represented by the cocycle

$$\langle \mathrm{Im}\,\hat{C}_2, [g_1|g_2|g_3] \rangle = \mathrm{Vol}(\infty, g_1\infty, g_1 g_2\infty, g_1 g_2 g_3\infty)/4\pi^2$$

Now for a hyperbolic simplex with vertices $(\infty, 0, 1, z)$ it is a classical calculation that

$$\mathrm{Vol}(\infty, 0, 1, z) = \mathcal{D}(z)$$

where $\mathcal{D}$ is the *Bloch-Wigner function* defined for all $z \in \mathbb{C} - \{0, 1\}$ by the formula

$$(10.9) \qquad \mathcal{D}(z) = \arg(1 - z)\log|z| - \mathrm{Im}\int_0^z \frac{\log(1 - t)}{t}\,dt.$$

(cf. [**Bloch, 1978**].) In this formula the integral is carried out along a path starting with the interval $[0, \frac{1}{2}]$, and the arg and log functions are branches along the same arc. Then the indeterminacy of the first term cancels with the one of the second, so that $\mathcal{D}(z)$ is well-defined. We note that

$$-\int_0^z \frac{\log(1 - t)}{t}\,dt = \mathrm{Li}_2(z) = \sum_{n=1}^\infty \frac{z^n}{n^2}, \quad |z| \le 1$$

is the *dilogarithmic function* introduced by Euler (cf. [**Lewin, 1981**].) We have thus proved

THEOREM 10.10. $\mathrm{Im}\,\hat{C}_2 \in H^3(\mathrm{Sl}(2, \mathbb{C}), \mathbb{R})$ *is represented by the cocycle*

$$\langle \mathrm{Im}\,\hat{C}_2, [g_1|g_2|g_3] \rangle = \mathcal{D}(\{\infty : g_1\infty : g_1 g_2\infty : g_1 g_2 g_3\infty\})/4\pi^2$$

We want to extend this formula to be valid also for the real part. This we do in terms of the exact sequence in theorem 8.19 for $F = \mathbb{C}$

by defining for $z \in \mathbb{C} - \{0, 1\}$ the following expression $\rho(z) \in \wedge_{\mathbb{Z}}^2(\mathbb{C})$ analogous to (10.9):

(10.11)
$$\rho(z) = \frac{\log z}{2\pi i} \wedge \frac{\log(1 - z)}{2\pi i} + 1 \wedge \frac{1}{(2\pi i)^2} \int_0^z \left(\frac{\log(1 - t)}{t} + \frac{\log t}{1 - t} \right) dt$$

Here again the indeterminacy of the first term cancels out with that of the second if the integral and the branches of the logarithms are taken along the same paths. We can now state (cf. [**Dupont, 1987**]):

THEOREM 10.12.   i) ρ *defines a well-defined map* $\mathcal{P}_{\mathbb{C}} \to \wedge_{\mathbb{Z}}^2(\mathbb{C})$.
ii) *The following diagram with exact rows commute:*

$$\begin{array}{ccccc}
H_3(\mathrm{Sl}(2, \mathbb{C})) & \xrightarrow{\sigma} & \mathcal{P}_{\mathbb{C}} & \xrightarrow{\lambda} & \wedge_{\mathbb{Z}}^2(\mathbb{C}^{\times}) \\
\downarrow{\scriptstyle 2\hat{C}_2} & & \downarrow{\scriptstyle \rho} & & \downarrow \\
0 \xrightarrow{} \mathbb{C}/\mathbb{Q} & \xrightarrow{1 \wedge \mathrm{id}} & \wedge_{\mathbb{Z}}^2(\mathbb{C}) & \xrightarrow{e \wedge e} & \wedge_{\mathbb{Z}}^2(\mathbb{C}^{\times})
\end{array}$$

where σ *and* λ *are given as in theorem 8.19 and* $e(v) = \exp(2\pi i v)$ *for* $v \in \mathbb{C}$.

Proof: We shall not give all details of this; but first we notice that theorem 10.10 shows that $\rho \circ \sigma$ and $(1 \wedge \mathrm{id}) \circ (2\hat{C}_2)$ agree on $H_3(\mathrm{Sl}(2, \mathbb{C}))^-$. In fact the inclusion $\mathbb{R} \to \wedge_{\mathbb{Z}}^2(\mathbb{C})^-$ sending $r \mapsto 1 \wedge ir$ is split by the map p^- given by $p^-(u \wedge v) = \frac{1}{2i}(\bar{u}v - \bar{v}u)$ and one checks that

$$p^- \circ \rho(z) = -(2\pi i)^{-2}(\mathcal{D}(z) - \mathcal{D}(1 - z)) = \mathcal{D}(z)/2\pi^2.$$

Hence by theorem 10.10

$$p^- \circ \rho(\sigma(z)) = \mathcal{D}(\sigma(z))/2\pi^2 = 2\langle \mathrm{Im}\,\hat{C}_2, z \rangle \quad \text{for } z \in H_3(\mathrm{Sl}(2, \mathbb{C})).$$

To show that $\rho \circ \sigma$ and $(1 \wedge \mathrm{id}) \circ (2\hat{C}_2)$ agree on $H_3(\mathrm{Sl}(2, \mathbb{C}))^+$ we next reduce the problem to $\mathrm{Sl}(2, \mathbb{R})$ by the following

LEMMA 10.13. *The inclusion* $\mathrm{Sl}(2, \mathbb{R}) \subseteq \mathrm{Sl}(2, \mathbb{C})$ *induces a surjection*

$$H_3(\mathrm{Sl}(2, \mathbb{R})) \longrightarrow H_3(\mathrm{Sl}(2, \mathbb{C}))^+$$

Remark: This map is also injective cf. [**Sah, 1989**].

Proof: By theorem 8.19 and a similar sequence for $\mathrm{Sl}(2,\mathbb{R})$ acting on $P^1(\mathbb{R})$ we get the following commutative diagram with exact rows:

$$
\begin{array}{ccccccc}
\to H_3(\mathrm{Sl}(2,\mathbb{R})) & \longrightarrow & \mathcal{P}_{\mathbb{R}} & \xrightarrow{\lambda_{\mathbb{R}}} & \wedge^2_{\mathbb{Z}}(\mathbb{R}^\times) & \longrightarrow & H_2(\mathrm{Sl}(2,\mathbb{R})) \to \\
\downarrow & & \downarrow & & \downarrow & & \downarrow \\
\to H_3(\mathrm{Sl}(2,\mathbb{C}))^+ & \longrightarrow & \mathcal{P}^+_{\mathbb{C}} & \xrightarrow{\lambda^+} & \wedge^2_{\mathbb{Z}}(\mathbb{C}^\times)^+ & \longrightarrow & H_2(\mathrm{Sl}(2,\mathbb{C}))^+ \to
\end{array}
$$

In this diagram

$$\wedge^2_{\mathbb{Z}}(\mathbb{C}^\times)^+ \cong \wedge^2_{\mathbb{Z}}(\mathbb{R}^\times) \oplus \wedge^2_{\mathbb{Z}}(\mathrm{U}(1))$$

and similarly λ^+ splits into

$$\lambda^+ = \lambda_{\mathbb{R}} \oplus \lambda_{\mathrm{U}(1)}$$

where $\lambda_{\mathrm{U}(1)}$ is given by

$$\lambda_{\mathrm{U}(1)}(z) = \frac{z}{|z|} \wedge \frac{1-z}{|1-z|}.$$

To show the surjectivity of the vertical map on the left it suffices to show that $\lambda_{\mathrm{U}(1)}$ induces an isomorphism

$$\mathcal{P}^+_{\mathbb{C}}/\mathcal{P}_{\mathbb{R}} \xrightarrow{\lambda_{\mathrm{U}(1)}} \wedge^2_{\mathbb{Z}}(\mathrm{U}(1))$$

This is simply done by constructing the inverse ν : For $e^{i\alpha}, e^{i\beta} \in \mathrm{U}(1)$ with $0 < \alpha, \beta < 2\pi$,

$$\nu(e^{i\alpha} \wedge e^{i\beta}) = \{z\}$$

where $z \in \mathbb{C}$ is the intersection of the two lines through 0 respectively 1 and making the angle α respectively β with the real line (see fig.).

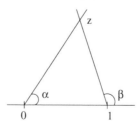

For $\alpha = \beta \pm \pi$, we put of course $\nu(e^{i\alpha} \wedge e^{i\beta}) = 0$. Clearly ν is alternating since in $\mathcal{P}_{\mathbb{C}}$

$$\nu(e^{i\beta} \wedge e^{i\alpha}) = \{1-z\} = -\{z\}.$$

It remains to show that ν is bi-additive in α and β. This requires another application of the identity in theorem 8.14 for a suitable rational function. We refer to [**Dupont-Parry-Sah, 1988**] for the details. $\square$

To finish the proof of theorem 10.12 we give a formula for $\hat{P}_1 = \hat{C}_2|H_3(\mathrm{Sl}(2,\mathbb{R}))$ in terms of $\mathcal{P}_\mathbb{R}$ and the following function:

$$L(x) = -\frac{\pi^2}{6} - \frac{1}{2}\int_0^x \left(\frac{\log t}{1-t} + \frac{\log(1-t)}{t}\right) dt$$

$$= -\frac{\pi^2}{6} + \sum_{n=0}^{\infty} \frac{x^n}{n^2} + \frac{1}{2}(\log x)(\log(1-x))$$

for $0 < x < 1$ and $L(0) = -\frac{\pi^2}{6}$, $L(1) = 0$. Without the constant term this function is called *Rogers' L-function*. It is a classical fact that this function satisfies the following identities (cf. [**Coxeter, 1935**], [**Lewin, 1981**]).

(10.14,i) $\qquad L(x) + L(1-x) = -\frac{\pi^2}{6} = L(0)$

(10.14,ii) $\quad L(x) - L(y) + L\left(\frac{y}{x}\right) - L\left(\frac{1-x^{-1}}{1-y^{-1}}\right) + L\left(\frac{1-x}{1-y}\right) = 0$

for $0 < y < x < 1$. Then we extend L to $\mathbb{R}_+$ by

$$L(x) = -L(1/x), \quad x > 1,$$

and use (10.14,i) to define $L(x)$ for $x < 0$. It is now clear from definition (8.13) that L induces a well-defined homomorphism $L : \mathcal{P}_\mathbb{R} \to \mathbb{R}/\mathbb{Z}(\pi^2)$ and we now have the following real version of theorem 10.12:

THEOREM 10.15. *The map* $(1/4\pi^2)L\circ\sigma : H_3(\mathrm{Sl}(2,\mathbb{R})) \to \mathbb{R}/\mathbb{Z}(1/4)$ *represents the reduction* mod $1/4$ *of the Cheeger-Chern-Simons class* $\hat{P}_1$ *associated to the first Pontrjagin polynomial.*

With this theorem it is rather straightforward to deduce theorem 10.12 in view of lemma 10.13. We refer to [**Dupont, 1987**] for the details of the proof. The main idea is to lift $\hat{P}_1$ to the universal covering $\mathrm{Sl}(2,\mathbb{R})^\sim$ where this class can be represented by a real valued cochain via proposition 10.4 and where we again integrate the volume form over some simplices in $\mathrm{Sl}(2,\mathbb{R})^\sim$ considered as a cylinder over the hyperbolic plane.

Remark 1.   We must reduce   mod $1/4$ in theorem 10.15 since σ : $H_3(\mathrm{Sl}(2,\mathbb{R})) \to \mathcal{P}_\mathbb{R}$ factors through $H_3(\mathrm{PSl}(2,\mathbb{R}))$ and

$$\ker\left(H_3(\mathrm{Sl}(2,\mathbb{R})) \to H_3(\mathrm{PSl}(2,\mathbb{R}))\right) = \mathbb{Z}/4.$$

It should however be noted that in fact L gives an explicit cochain representing $\tilde{P}_1 \in H_3(\mathrm{Sl}(2,\mathbb{R})^{\sim}, \mathbb{R})$, where $\hat{P}_1$ is the unique lift of $\hat{P}_1$ corresponding to a *continuous cochain* on $\mathrm{Sl}(2,\mathbb{R})^{\sim}$. For a discussion of this see [**Dupont-Sah, 1994**], which also contains a homological proof of some interesting identities involving the function L which has occurred in connection with conformal field theory.

Remark 2.   Theorem 10.12 only gives a formula for $\hat{C}_2$ with $\mathbb{C}/\mathbb{Q}$ coefficients. However for the restriction $\hat{P}_1$ to $\mathrm{Sl}(2,\mathbb{R})$ theorem 10.15 gives an explicit lift with $\mathbb{R}/\mathbb{Z}\left(\frac{1}{4}\right)$ coefficients (cf. also remark 1 above). A similar direct lift for $\hat{C}_2$ has been given by [**Neumann-Yang, 1999**].

Remark 3.   Theorem 10.12 is the analogue of the tautological formula for $\hat{C}_1 \in H^1(\mathrm{Gl}(1,\mathbb{C}), \mathbb{C}/\mathbb{Z})$

$$\hat{C}_1(g) = \frac{1}{2\pi}\log(g), \quad g \in \mathrm{Gl}(1,\mathbb{C}) = \mathbb{C}^*.$$

The formula for $\mathrm{Im}\,\hat{C}_2$ in theorem 10.10 is due to Bloch and Wigner (unpublished cf. [**Bloch, 1978**]). Similar formulas for $\mathrm{Im}\,\hat{C}_3$ has been obtained by [**Goncharov, 1995**] and [**Hain-Yang, 1996**].

Returning to the general Cheeger-Chern-Simons class as defined by (10.6)

$$\hat{C}_k = [\mathcal{J}(\omega_k)] \in H^{2k-1}(\mathrm{Gl}(n,\mathbb{C}), \mathbb{C}/\mathbb{Z}) = \mathrm{Hom}(H_{2k-1}, (\mathrm{Gl}(n,\mathbb{C})), \mathbb{C}/\mathbb{Z})$$

we notice that it is given by integrating the form ω_k over some simplices. In general this gives transcendental functions in the variables of the vertices similarly to the dilogarithmic function considered above. However $H_*(\mathrm{Gl}(n,\mathbb{C}))$ is a purely algebraic object on which $\alpha \in \mathrm{Gal}(\mathbb{C}/\mathbb{Q})$, the group of field automorphisms, acts by

$$\alpha_*[g_1|\ldots|g_q] = [\alpha g_1|\ldots|\alpha g_q], \quad g_i \in \mathrm{Gl}(n,\mathbb{C}).$$

Hence we get à priori new invariants by composing $\hat{C}_k$ with α (cf. [**Cheeger, 1974**]). That is, we define $\hat{C}_k^{\prime\alpha}$ the *twisted* Cheeger-Chern-Simons invariant by

$$\langle \hat{C}_k^\alpha, [g_1|\ldots|g_{2k-1}]\rangle = \langle \hat{C}_k, [\alpha g_1|\ldots|\alpha g_{2k-1}]\rangle, \quad g_i \in \mathrm{Gl}(n,\mathbb{C}).$$

This procedure has been used very successfully by A. Borel in his famous calculation [**Borel, 1977**] of $H_*(\mathrm{Sl}(K,\infty),\mathbb{Q})$ for $K \subset \mathbb{C}$ an algebraic number field, where

$$\mathrm{Sl}(\infty, K) = \lim_{n\to\infty} \mathrm{Sl}(n,K)$$

under the natural inclusions $\mathrm{Sl}(n,K) \subseteq \mathrm{Sl}(n+1,K)$. Borel actually only needed the Galois-twisted r_k^α, $\alpha \in \mathrm{Gal}(\mathbb{C}/\mathbb{Q})$, where r_k is the Borel *regulator* given by

(10.16) $$r_k = \mathrm{Im}\,\hat{C}_k \in H^{2k-1}(\mathrm{Gl}(n,\mathbb{C}),\mathbb{R}).$$

Notice that since $K/\mathbb{Q}$ is assumed to be a finite extension there is some other finite extension $L \supset K$ containing all the fields $\alpha(K), \alpha \in \mathrm{Gal}(\mathbb{C}/\mathbb{Q})$. Hence the restriction of $\hat{C}_k^\alpha$ and r_k^α only depends on the image of α in the finite group $\mathrm{Gal}(L/\mathbb{Q})$. For $k_1 < k_2 < \cdots < k_l$ and $\alpha_1,\ldots,\alpha_l \in \mathrm{Gal}(L/\mathbb{Q})$ we therefore obtain the product of the cohomology classes

$$r_{k_1\ldots k_l}^{\alpha_1\ldots\alpha_l} = r_{k_1}^{\alpha_1} \otimes \cdots \otimes r_{k_l}^{\alpha_l} \in H^{2(k_1+\cdots+k_l)-l}(\mathrm{Sl}(n,K),\mathbb{R}^{\otimes l})$$

$(\mathbb{R}^{\otimes l} = \mathbb{R} \otimes_{\mathbb{Z}} \cdots \otimes_{\mathbb{Z}} \mathbb{R})$.
In particular for the algebraic closure $\bar{\mathbb{Q}}$ of the rationals, i.e. $\bar{\mathbb{Q}} = \lim_{\overrightarrow{K}} K$, with $[K : \mathbb{Q}] < \infty$ the above set of invariants gives a homomorphism, the *regulator map*:

(10.17)
$$\mathbf{r} : H_*(\mathrm{Sl}(\infty,\bar{\mathbb{Q}})) \to \lim_{\overrightarrow{K}} \prod_{l,k_1<\cdots<k_l} \mathrm{Hom}(\mathbb{Z}[\mathrm{Gal}(K/\mathbb{Q})^l],\mathbb{R}^{\otimes l})$$
$$\mathbf{r} = (r_{k_1\ldots k_l}^{\alpha_1\ldots\alpha_l}(z)), \quad z \in H_*(\mathrm{Sl}(\infty,\bar{\mathbb{Q}})).$$

We can now state the following consequence of Borel's theorem (the original is much more precise):

THEOREM 10.18. *The regulator map* $\mathbf{r}$ *induces an injective map on* $H_*(\mathrm{Sl}(\infty,\bar{\mathbb{Q}}),\mathbb{Q})$.

Remark: Since by Suslin's result (theorem 9.22 and the following remark 1) the torsion is determined by the topological classifying space $B\,\mathrm{Sl}(\infty, \mathbb{C})$, we therefore have complete invariants to detect $H_*(\mathrm{Sl}(\infty, \bar{\mathbb{Q}}))$.

Let us mention an application of Borel's theorem 10.18 to hyperbolic geometry. As in remark 2 at the end of chapter 8 an oriented compact hyperbolic manifold has the form $M = \Gamma \backslash \mathcal{H}^n$ for a co-compact discrete subgroup $\Gamma \subseteq \mathrm{SO}(1, n)$, and thus determines a fundamental class $[M] \in H_n(\mathrm{SO}(1, n))$. Also it is a well-known consequence of the Weil rigidity theorem that Γ can be chosen algebraic, i.e. inside $\mathrm{SO}(1, n, \bar{\mathbb{Q}} \cap \mathbb{R})$ (cf. [**Raghunatan, 1972**, chap. VI, prop. 6.6]). By the inclusion

$$\mathrm{SO}(1, n, \bar{\mathbb{Q}} \cap \mathbb{R}) \subset \mathrm{Sl}(n + 1, \bar{\mathbb{Q}} \cap \mathbb{R}) \subset \mathrm{Sl}(n + 1, \bar{\mathbb{Q}})$$

we can evaluate our characteristic classes on $[M]$ and in particular the image of $[M]$ in $H_n(\mathrm{Sl}(\infty, \bar{\mathbb{Q}}), \mathbb{Q})$ is determined by the regulators

$$r_{k_1 \cdots k_l}^{\alpha_1 \cdots \alpha_l}(M) = \langle r_{k_1 \cdots k_l}^{\alpha_1 \cdots \alpha_l}, [M] \rangle \in \mathbb{R}^{\otimes l}$$

for $2(k_1 + \cdots + k_l) - l = n$. Notice that for $n = 4k - 1$ we have $r_{2k}(M) = 0$ and it is shown in [**Dupont-Kamber, 1993**, cor. 5.7] (cf. [**Yoshida, 1985**], [**Neumann-Zagier, 1985**]) that

$$\eta(M) \equiv (-1)^k \lambda_k \langle \hat{C}_{2k}, [M] \rangle \quad \mathrm{mod}\ \mathbb{Q}$$

where $\eta(M)$ is the Atiyah-Patodi-Singer η-invariant and

$$\lambda_k = 2^{2k}(2^{2k-1} - 1)B_k / (2k)!$$

with B_k the kth Bernoulli number (cf. [**Hirzebruch, 1966**, chap. I, sect. 1.5]). Hence we obtain the following curious result:

COROLLARY 10.19. *If two compact oriented hyperbolic manifolds M_1 and M_2 of dimension $n = 4k - 1$ have equal regulators, i.e. if*

$$r_{k_1 \cdots k_l}^{\alpha_1 \cdots \alpha_l}(M_1) = r_{k_1 \cdots k_l}^{\alpha_1 \cdots \alpha_l}(M_2)$$

for all $k_i > 1$ and $\alpha_i \in \mathrm{Gal}(\mathbb{C}/\mathbb{Q})$ then

$$\eta(M_1) \equiv \eta(M_2) \quad \mathrm{mod}\ \mathbb{Q}.$$

Remark: Notice that if $M_j, j = 1, 2$, are defined over a number field K such that one of the Galois transformations α_i satisfies $\alpha_i(K \cap \mathbb{R}) \subseteq \mathbb{R}$ then $r_{k_1 \cdots k_2}^{\alpha_1 \cdots \alpha_l}(M_j) = 0$. Hence the condition is trivially fulfilled unless *all* α_i are *not* of this type.

Let us also mention without proof another remarkable geometric result on the Cheeger-Chern-Simons classes, the *Bloch conjecture* proved by [**Reznikov, 1995**]:

THEOREM 10.20. *If* $\Gamma = \pi_1(X)$, X *a smooth complex projective variety and* $\varphi : \Gamma \to \mathrm{Sl}(n, \mathbb{C})$ *a representation, then* $\varphi^* \hat{C}_k = 0$ *in* $H^{2k-1}(\Gamma, \mathbb{C}/\mathbb{Q})$.

For 3-dimensional scissors congruences Borel's theorem is particularly useful in degrees ≤ 3: Here there is only one regulator namely r_2 since $r_1 = 0$ when restricted to Sl (by definition). In this case we conclude from theorem 10.18 and corollary 9.23 (or rather the corresponding result of Suslin for $\bar{\mathbb{Q}}$ instead of $\mathbb{C}$):

COROLLARY 10.21.   i) $H_2(\mathrm{Sl}(2, \bar{\mathbb{Q}})) = 0$.
ii) *A complete set of invariants for* $H_3(\mathrm{Sl}(2, \bar{\mathbb{Q}}))$ *is given by the twisted Borel regulators* $\{r_2^\alpha; \alpha \in \mathrm{Gal}(\bar{\mathbb{Q}}/\mathbb{Q})\}$ *together with the Cheeger-Chern-Simons invariant* $\mathrm{Re}\,\hat{C}_2$.

Proof: Since by (9.5) and corollary 8.20, $H_2(\mathrm{Sl}(2, \bar{\mathbb{Q}})) = H_2(\mathrm{Sl}(\infty, \bar{\mathbb{Q}}))$ is a $\mathbb{Q}$-vector space and since r_2 is a 3-dimensional class i) clearly follows. Similarly by (9.6) and i) above $H_3(\mathrm{Sl}(2, \bar{\mathbb{Q}})) = H_3(\mathrm{Sl}(\infty, \bar{\mathbb{Q}}))$ and by corollary 8.20 and theorem 10.18 it remains to show that the torsion group $H_3(\mu_{\bar{\mathbb{Q}}}) = \mathbb{Q}/\mathbb{Z}$ is detected by $\mathrm{Re}\,\hat{C}_2$. But this clearly follows from the formula in theorem 10.2 since the image by σ of $1/q \in \mathbb{Q}/\mathbb{Z}$ corresponds to the rational lune $L(2\pi/q)$ (cf. remarks 2 and 3 at the end of chapter 7). $\qquad \square$

Let us summarize the consequences for scissors congruences: Suppose that P_1, P_2 are two spherical respectively two hyperbolic polyhedra such that $D(P_1) = D(P_2)$. Then by the sequences (10.1,$\pm$) we obtain a unique "difference element" $d(P_1, P_2) \in H_3(\mathrm{Sl}(2, \mathbb{C}))^\pm$ such that

$$\sigma(d(P_1, P_2)) = [P_1] - [P_2]$$

in $\mathcal{P}(S^3)/\mathbb{Z}$ respectively $\mathcal{P}(\mathcal{H}^3)$. Hence we have the following:

COROLLARY 10.22. *Let* P_1 *and* P_2 *be two spherical respectively two hyperbolic polyhedra with all vertices defined over* $\bar{\mathbb{Q}}$, *the field of algebraic numbers. Then* P_1 *and* P_2 *are scissors congruent if and only if*

   i) $D(P_1) = D(P_2)$

   ii) $\mathrm{Vol}(P_1) = \mathrm{Vol}(P_2)$

   iii) $r_2^\alpha(d(P_1, P_2)) = 0 \quad \forall \alpha \in \mathrm{Gal}(\bar{\mathbb{Q}}/\mathbb{Q}), \quad \alpha \neq \mathrm{id}$.

Remark 1. We have excluded $\alpha = \mathrm{id}$ in (iii) since it is listed separately in (ii). Note that even in the spherical case only hyperbolic volumes enters in (iii). In fact by theorem 10.2, r_2^α is given by

$$r_2^\alpha(z) = \frac{1}{4\pi^2} \mathrm{Vol}_{\mathcal{H}^3} \circ \sigma((\alpha_* z)_-) \quad z \in H_3(\mathrm{Sl}(2, \mathbb{C}))$$

where

$$\alpha_* z = (\alpha_* z)_+ + (\alpha_* z)_-$$

is the decomposition in the ± 1-eigenspaces in (10.1,$\pm$). The spherical volume function is only used to distinguish the rational lunes.

Remark 2. As mentioned above a 3-dimensional compact hyperbolic manifold has a presentation as $\Gamma \backslash \mathcal{H}^3$ for $\Gamma \subset \mathrm{Sl}(2, \mathbb{C})$ a lattice defined over an algebraic number field. Hence the fundamental domain for such a lattice is an example of a polyhedron with zero Dehn invariant for which corollary 10.22 applies. We shall give more examples in the next chapter.

Returning to the Galois action we note that this does not make sense in general for a polyhedron. This is the reason why we had to express the Galois action in corollary 10.22 using the "difference element". However the Galois action clearly makes sense in $\mathcal{P}_{\mathbb{C}}$ by

$$\alpha_*\{z\} = \{\alpha z\} \quad \text{for } z \in \mathbb{C} - \{0, 1\}, \quad \alpha \in \mathrm{Gal}(\mathbb{C}/\mathbb{Q}).$$

But unless α commutes with complex conjugation it does not necessarily keep the (± 1)-eigenspaces invariant. We shall see examples of that in the next chapter. Here we just notice that we get an extension of $2\hat{C}_2^{\prime \alpha}$ to $\mathcal{P}_{\mathbb{C}}$ by defining for $\alpha \in \mathrm{Gal}(\mathbb{C}/\mathbb{Q})$

(10.23) $\rho^\alpha(z) = \rho(\alpha(z)), \qquad z \in \mathbb{C} - \{0, 1\}.$

Similarly we have $\mathcal{D}^\alpha : \mathcal{P}_{\mathbb{C}} \to \mathbb{R}$ defined by

$$\mathcal{D}^\alpha(z) = \mathcal{D}(\alpha(z)), \qquad z \in \mathbb{C} - \{0, 1\},$$

and we conclude:

THEOREM 10.24.   a) *Let $z_1, \ldots, z_l \in \bar{\mathbb{Q}} - \{0, 1\}$. Then the relation*

$$\sum_i \{z_i\} = 0$$

holds in $\mathcal{P}_{\bar{\mathbb{Q}}}$, i.e. is a consequence of the relation (8.13), if and only if

(i) $\sum_i \lambda\{z_i\} = 0$

(ii) $\sum_i \mathcal{D}(\alpha z_i) = 0 \quad \forall \alpha \in \mathrm{Gal}(\bar{\mathbb{Q}}/\mathbb{Q}).$

b) *Furthermore if all z_i are real algebraic numbers then the relation $\sum_i \{z_i\} = 0$ holds in $\mathcal{P}_{\mathbb{R}}$ if and only if (i) and (ii) hold together with*

(iii) $\sum_i L(z_i) \equiv 0 \qquad \mathrm{mod}\ \pi^2.$

Remark 1.   We use here that $H_3(\mathrm{Sl}(2, \mathbb{R})) \to H_3(\mathrm{Sl}(2, \mathbb{C}))$ is injective cf. [**Sah, 1989**]. Again (iii) is only needed to determine the $\mathbb{Q}/\mathbb{Z}$ component. Also in the real case there is another version with L lifted to $\mathbb{R}$ in (iii) (cf. remark 1 following theorem 10.15).

Remark 2.   If in theorem 10.24 all z_i and all their Galois conjugates αz_i are *real* then (ii) is clearly fulfilled. Hence already (i) implies that $\sum_i \{z_i\} = 0$ holds in $\mathcal{P}_{\mathbb{R}}$ modulo the torsion group. In particular we immediately get in this case

$$\sum_i L(z_i) \equiv 0 \quad \mathrm{mod}\ \mathbb{Q}\pi^2.$$

EXAMPLE 10.25. Let $\sigma = \frac{1}{2}(\sqrt{5} - 1)$, then we claim that

$$\{\sigma^{20}\} \equiv 2\{\sigma^{10}\} + 15\{\sigma^4\} \quad \text{in } \mathcal{P}_{\mathbb{R}} \quad \text{mod torsion}.$$

By the remark 2 above it suffices to show that

$$\lambda(\sigma^{20}) = 2\lambda(\sigma^{10}) + 15\lambda(\sigma^4) \quad \text{in } \wedge_{\mathbb{Z}}^2 (\mathbb{R}^\times).$$

For this one first shows by iterated use of the equation $\sigma^2 + \sigma - 1 = 0$ that

$$\lambda(-\sigma^{10}) = 15\lambda(-\sigma^2).$$

Hence

$$\lambda(\sigma^{20}) = 20\sigma \wedge (1 - \sigma^{10}) + 20\sigma \wedge (1 + \sigma^{10}) = 2\lambda(\sigma^{10}) + 30\lambda(-\sigma^2)$$
$$= 2\lambda(\sigma^{10}) + 15\lambda(\sigma^4)$$

since $\lambda(\sigma^2) = 2\sigma \wedge (1 - \sigma^2) = 2\sigma \wedge \sigma = 0$. In particular we have proved

$$15L(\sigma^4) + 2L(\sigma^{10}) - L(\sigma^{20}) \equiv 0 \quad \bmod \mathbb{Q}\pi^2.$$

It is a classical result of [**Coxeter, 1935**] that this multiple of π^2 is $\frac{7}{15} - \frac{16}{6} = -\frac{33}{15}$. The above identity in $\mathcal{P}_\mathbb{C}$ can also be proved using Rogers' identities (theorem 8.14).

Remark 3.   One could hope that the invariants ρ^α in (10.23) would detect all of $\mathcal{P}_\mathbb{C}$. However as we shall see in chapter 12 the set of these invariants does not give more information than the set of $\mathcal{D}^\alpha$'s. Hence the question is really if the natural map $H_3(\mathrm{Sl}(2, \bar{\mathbb{Q}})) \to H_3(\mathrm{Sl}(2, \mathbb{C}))$ is an isomorphism. This map is known to be injective; the surjectivity is often called the "Rigidity Conjecture" (cf. [**Sah, 1989**]).

CHAPTER 11

Simplices in spherical and hyperbolic 3-space

As mentioned in the previous chapter it does not make sense in general to take the Galois transformed of a spherical or hyperbolic polytope. On the other hand a Galois transformation $\tau \in \mathrm{Gal}(\mathbb{C}/\mathbb{Q})$ clearly induces an automorphism τ_* of $H_*(\mathrm{Sl}(2,\mathbb{C}))$ or $\mathcal{P}_\mathbb{C}$. In this chapter we shall consider a simplex in spherical or hyperbolic space and in favourable cases τ applied to the coordinates of the vertices will result in another simplex in one of these geometries - but not necessarily in the same geometry as that of the given simplex.

In order to ensure that this Galois action is compatible with the induced action on the appropriate homology group we consider the geometry of the complex sphere

$$S_\mathbb{C}^n = \{z \in \mathbb{C}^{n+1} \mid z_0^2 + z_1^2 + \cdots + z_n^2 = 1\}$$

with the action of the orthogonal groups $O(n+1, \mathbb{C})$, and we observe that

$$S^n = S_\mathbb{C}^n \cap \mathbb{R}^{n+1}, \quad \mathcal{H}^n = S_\mathbb{C}^n \cap (\mathbb{R}_+ \oplus i\mathbb{R}^n).$$

Similarly to the construction in chapter 3 we again have a Tits-complex $\mathcal{T}(S_\mathbb{C}^n)$ of *non-degenerate* subspaces. That is, we require the subspaces to be of the form $U = S_\mathbb{C}^n \cap V$, $V \subseteq \mathbb{C}^{n+1}$ a linear subspace which is non-degenerate for the quadratic form. Again with

$$\mathrm{St}(S_\mathbb{C}^n) = \tilde{H}_{n-1}(\mathcal{T}(S_\mathbb{C}^n), \mathbb{Z})$$

we get a Lusztig exact sequence

$$(11.1) \quad 0 \to \mathrm{St}(S_\mathbb{C}^n) \to \bigoplus_{U^{n-1}} \mathrm{St}(U^{n-1}) \to \cdots \to \bigoplus_{U^0} \mathrm{St}(U^0) \to \mathbb{Z} \to 0.$$

We now simply define

$$\mathcal{P}(S_\mathbb{C}^n) = H_0(O(n+1, \mathbb{C}), \mathrm{St}(S_\mathbb{C}^n)^t)$$

and similarly to the computation in example 7.11 we obtain in the case $n = 3$ the following exact sequences

(11.2)
$$0 \to \mathbb{Q}/\mathbb{Z} \to H_3(O(4,\mathbb{C}), \mathbb{Z}^t) \to \mathcal{P}(S^3_\mathbb{C}) \to \mathbb{C}^\times \otimes \mathbb{C}^\times \to$$
$$\to H_2(O(4,\mathbb{C}), \mathbb{Z}^t) \to 0.$$

Again since $\mathrm{Spin}(4,\mathbb{C}) \cong \mathrm{Sl}(2,\mathbb{C}) \times \mathrm{Sl}(2,\mathbb{C})$ and the $\mathbb{Z}/2$ factor in $\mathrm{Pin}(4,\mathbb{C})$ acts by permuting the factors we get by Künneth's theorem

$$H_i(O(4,\mathbb{C}), \mathbb{Z}^t) \cong H_i(\mathrm{Sl}(2,\mathbb{C})), \quad i \le 3.$$

Now using the isomorphisms (theorem 7.4,a) and (2.14))

$$\mathcal{P}(S^3)/\Sigma\mathcal{P}(S^2) \cong H_0(O(4), \mathrm{St}(S^3)^t)$$
$$\mathcal{P}(\mathcal{H}^3) \cong H_0(O^1(1,3), \mathrm{St}(\mathcal{H}^3)^t)$$

and comparing with our sequences (10.1,$\pm$) we get the following commutative diagram with exact rows and with injective maps in the columns:

(11.3)
$$\begin{array}{ccccc}
0 \to \mathbb{Q}/\mathbb{Z} & \longrightarrow & H_3(\mathrm{Sl}(2,\mathbb{C}))^+ & \xrightarrow{\sigma} & \mathcal{P}(S^3)/\Sigma\mathcal{P}(S^2) \\
\downarrow & & \downarrow & & \downarrow \\
0 \to \mathbb{Q}/\mathbb{Z} & \longrightarrow & H_3(\mathrm{Sl}(2,\mathbb{C})) & \xrightarrow{\sigma} & \mathcal{P}(S^3_\mathbb{C}) \\
\uparrow & & \uparrow & & \uparrow \\
0 & \longrightarrow & H_3(\mathrm{Sl}(2,\mathbb{C}))^- & \xrightarrow{\sigma} & \mathcal{P}(\mathcal{H}^3)
\end{array}$$

Thus a Galois transformation τ induces an endomorphism of the groups in the middle line of the diagram (11.3). Notice that the group $\mathbb{Q}/\mathbb{Z} \cong \mu_\mathbb{C}$ is the multiplicative group of unities but the induced action by τ is given by τ^2. Notice also that the inclusions on the right side of the diagram are induced by the inclusions $\mathbb{R}^4 \subseteq \mathbb{C}^4$ respectively $\mathbb{R} \oplus i\mathbb{R}^3 \subseteq \mathbb{C}^4$ or the other real subspaces obtained from these by the action of the group $O(4,\mathbb{C})$. The Galois transformation τ can of course map an element in the image from $\mathcal{P}(S^3)$ into another such element or map it into an element in the image from $\mathcal{P}(\mathcal{H}^3)$ or vice versa. But it may very well happen that the Galois transformed of a spherical or hyperbolic polyhedron is neither in the image from the top or the bottom in the diagram above. We shall study this more closely in the case of simplices using their *Gram matrices*. Let us briefly review this parametrisation of the set of simplices in spherical and hyperbolic geometry of any dimension

$n \geq 1$. There is a similar construction in the Euclidean geometry, but we shall not discuss that case.

For $\varepsilon = \pm 1$ our model for the spherical respectively hyperbolic geometry is the conic surface in $\mathbb{R}^{n+1}$

$$S_{\varepsilon}^n = \{x \in \mathbb{R}^{n+1} \mid \langle x, x \rangle_{\varepsilon} = \varepsilon, \text{ and } x_0 > 0 \text{ if } \varepsilon = -1\}$$

where $\langle \cdot, \cdot \rangle_{\varepsilon}$ is the symmetric bilinear form

$$\langle x, y \rangle_{\varepsilon} = \varepsilon x_0 y_0 + x_1 y_1 + \cdots + x_n y_n, \quad x, y \in \mathbb{R}^{n+1}.$$

In the *spherical* case the vertices of a simplex Δ is a set $(v_0, \ldots, v_n)$ of $n + 1$ $\mathbb{R}$-linearly independent unit vectors. We can then define, in a unique way, a *dual* simplex Δ^* with vertices $(u_0, \ldots, u_n)$ according to the following rules

(11.4) $\langle u_i, v_j \rangle_+ = 0, \quad i \neq j, \quad \langle u_i, v_i \rangle_+ < 0, \quad \langle u_i, u_i \rangle_+ = 1.$

The spherical simplex Δ is then given by the set

$$\Delta = \{v \in \mathbb{R}^{n+1} \mid \langle v, v \rangle_+ = 1, \langle v, u_i \rangle_+ \leq 0, \quad i = 0, \ldots, n\}.$$

The $(n + 1) \times (n + 1)$ matrix $G(\Delta) = (\langle u_i, u_j \rangle_+)$ is called the *Gram matrix* associated to the spherical n-simplex Δ, and it has the following properties:

(11.5) $\Delta = (\Delta^*)^*$ and $G(\Delta^*) = (\langle v_i, v_j \rangle_+)$.
(11.6) $G(\Delta)$ is a real symmetric $(n + 1) \times (n + 1)$-matrix.
(11.7) $G(\Delta)$ has diagonal entries equal to 1.
(11.8) For $0 \leq i \neq j \leq n + 1$ the (i, j)-th entry of $G(\Delta)$ is equal to $- \cos \theta_{ij}$ where $\theta_{ij} = $ the interior dihedral angle between the codimensional 1 faces opposite (v_i, v_j).
(11.9) Permutation of the vertices of Δ corresponds to simultaneous column and row permutations of $G(\Delta)$.
(11.10) The $n \times n$ principal minors of $G(\Delta)$ are positive definite.

(11.11,+) $\det G(\Delta) > 0.$

In the case of Δ a *hyperbolic* n-simplex the vertices $(v_0, \ldots, v_n)$ satisfies

(11.12) $\langle v_i, v_i \rangle_- = -1 \quad$ and $v_{i0} = -\langle v_i, e_0 \rangle_- > 0.$

Again the *dual* simplex Δ^* with vertices $(u_0, \ldots, u_n)$ is given by (11.4) with $\langle -, - \rangle_+$ replaced by $\langle -, - \rangle_-$ and again

$$\Delta = \{ v \in \mathbb{R}^{n+1} \mid \langle v, v \rangle_- = -1, \quad \langle v, u_i \rangle_- \le 0, \quad i = 0, \ldots, n \}$$

Also the *Gram matrix* $G(\Delta) = (\langle u_i, u_j \rangle_-)$ has the properties (11.5)-(11.10) whereas (11.11,+) is replaced by

(11.11,−) $\det G(\Delta) < 0.$

Note however that Δ^* in this case is *not* a hyperbolic simplex since the vertices u_i satisfy $\langle u_i, u_i \rangle_- = 1$ and thus are "ultra infinite".

PROPOSITION 11.13, +. *Suppose $G = (g_{ij})$ is a real $(n+1) \times (n+1)$-matrix satisfying (11.6), (11.7), (11.10) and (11.11,+). Then there is a spherical n-simplex Δ, unique up to isometry with $G(\Delta) = G$.*

Proof: The matrix G defines an inner product on an abstract $\mathbb{R}$-vector space of dimension $n + 1$ with basis $\{u_0, \ldots, u_n\}$. By change of basis this is isomorphic to $\mathbb{R}^{n+1}$ with the inner product $\langle \cdot, \cdot \rangle_+$ since G is positive definite. Now the vertices $(v_0, \ldots, v_n)$ of Δ are determined by the condition (11.4). $\square$

Remark: Given a spherical simplex Δ with vertices $(v_0, \ldots, v_n)$ we can decompose the n-sphere into 2^{n+1} simplices with vertices

$$(\varepsilon_0 v_0, \varepsilon_1 v_1, \ldots \varepsilon_n v_n) \quad \text{where} \quad \varepsilon_i = \pm 1.$$

The corresponding Gram matrix for each of these is obtained from $G(\Delta)$ through simultaneous multiplication of the ith column and ith row by ε_i. Notice that for $\Delta_i' = |(v_0, \ldots, -v_i, \ldots, v_n)|$ the union $\Delta \amalg \Delta_i'$ form a *suspension* (cf. chapter 3) of the $(n-1)$-simplex in the orthogonal complement of span$\{v_i\}$ whose dual $(n-1)$-simplex has vertices $(u_0, \ldots, \hat{u}_i, \ldots, u_n)$.

In the hyperbolic case we have:

PROPOSITION 11.13, −. *Suppose $G = (g_{ij})$ is a real $(n+1) \times (n+1)$-matrix satisfying (11.6), (11.7), (11.10) and (11.11,−). Then there is a set of signs $(\varepsilon_0, \ldots, \varepsilon_n), \varepsilon_i = \pm 1$, unique up to simultaneous multiplication by ± 1, and a hyperbolic n-simplex Δ, unique up to isometry, such that $G(\Delta) = (\varepsilon_i g_{ij} \varepsilon_j)$.*

Proof: We proceed exactly as in the proof of proposition 11.13,+. However in order to ensure condition (11.12) we may have to replace u_i by $-u_i$ thus multiplying the ith row and ith column of the Gram matrix by -1. □

Remark 1. In view of proposition 11.13, − and the remark following proposition 11.13,+ we will say that two symmetric matrices $G = (g_{ij})$ and $G' = (g'_{ij})$ are "equivalent modulo (even) sign changes" if for some vector $(\varepsilon_0, \ldots, \varepsilon_n), \varepsilon_i = \pm 1, (\prod_{i=0}^n \varepsilon_i = 1)$,

$$g_{ij} = \varepsilon_i g'_{ij} \varepsilon_j \qquad \forall i, j.$$

We have thus shown that the equivalence class modulo *even* sign changes of a real $(n+1) \times (n+1)$-matrix G satisfying (11.6), (11.7), (11.10) and (11.11,+) determines a well-defined element $[\Delta] \in \mathcal{P}(S^n)/\Sigma\mathcal{P}(S^{n-1})$ and an *odd* sign change replaces $[\Delta]$ by $-[\Delta]$. If G satisfies (11.6), (11.7), (11.10) and (11.11,−) there is a unique hyperbolic simplex Δ given as in proposition 11.13, − and we associate to G the element $\varepsilon[\Delta] \in \mathcal{P}(\mathcal{H}^n)$ where $\varepsilon = \prod_i \varepsilon_i$.

Remark 2. If we allow the hyperbolic simplex Δ to have some (or all) vertices on the boundary $\partial \mathcal{H}^n$ then the associated Gram matrix satisfies conditions (11.5)-(11.9), (11.11,−) and the following condition in place of (11.10):

$(11.10, -)$

All $(n+1) \times (n+1)$ principal minors are positive definite, and all $n \times n$ principal minors have non-negative determinant.

Again proposition 11.13, − holds in $\bar{\mathcal{H}}^n$ if we replace condition (11.10) with $(11.10, -)$.

In the following we assume $n = 3$. Returning to our diagram (11.3) it follows from the remark 1 above that up to even sign changes a 4×4 real matrix G satisfying (11.6), (11.7),(11.10) and (11.11,+) or (11.11,−) determines in both cases a well-defined element $\varepsilon[\Delta] \in \mathcal{P}(S_{\mathbb{C}}^3), \varepsilon = \pm 1$, where Δ is a spherical or hyperbolic 3-simplex. Hence if $\tau \in \mathrm{Gal}(\mathbb{C}/\mathbb{Q})$ is a field automorphism then $\varepsilon \tau_*[\Delta] \in \mathcal{P}(S_{\mathbb{C}}^3)$ is also well-defined, but the corresponding matrix τG need *not* be the Gram matrix of a spherical or hyperbolic simplex. On the other hand if it *is* a Gram matrix, i.e. satisfies our conditions above then the associated element in $\mathcal{P}(S_{\mathbb{C}}^3)$ is clearly $\varepsilon\tau_*[\Delta]$. Notice that the Galois action is compatible with the sign changes. Let us summarize our discussion in the following:

THEOREM 11.14. a) *Let G be a real 4×4-matrix and $\tau \in \mathrm{Gal}(\mathbb{C}/\mathbb{Q})$, and suppose that both G and τG are spherical or hyperbolic Gram matrices, i.e. satisfy (11.6), (11.7), (11.10) and $\det G \neq 0$. If $\varepsilon[\Delta]$ and $\varepsilon^\tau[\Delta^\tau]$ denote the corresponding elements in $\mathcal{P}(S^3_\mathbb{C})$ given by the propositions (11.13,$\pm$), then*

$$\tau_*(\varepsilon[\Delta]) = \varepsilon^\tau[\Delta^\tau].$$

b) *In particular, suppose for $z \in H_3(\mathrm{Sl}(2,\mathbb{C}))^\pm$ that the image $\sigma(z) \in \mathcal{P}(S^3)/\Sigma\mathcal{P}(S^2)$ (respectively $\mathcal{P}(\mathcal{H}^3)$) is of the form*

$$\sigma(z) = \sum_i \varepsilon_i[\Delta_i]$$

with each Δ_i given as in a). *Then*

$$\sigma(\tau_*(z)) = \sum_i \varepsilon_i^\tau[\Delta_i^\tau]$$

where each term is an element in $\mathcal{P}(S^3)/\Sigma\mathcal{P}(S^2)$ or $\mathcal{P}(\mathcal{H}^3)$.

$\square$

Remark: Note that in b) each simplex Δ_i^τ may be spherical or hyperbolic (depending on the sign of the corresponding Gram matrix) independently of the type of Δ_i.

DEFINITION 11.15. A (spherical or hyperbolic) n-simplex Δ is called *rational* if all its dihedral angles θ_{ij} are in $\mathbb{Q}\pi$.

Remark: Notice that if $\theta \in \mathbb{Q}\pi$ then $\cos\theta = (\zeta + \zeta^{-1})/2$ for ζ an mth root of unity for some m. Hence for any $\tau \in \mathrm{Gal}(\mathbb{C}/\mathbb{Q})$,

$$\tau(\cos\theta) = (\tau(\zeta) + \tau(\zeta)^{-1})/2,$$

where $\tau(\zeta)$ is again an m-th root of unity. Hence if G is the Gram matrix of a *rational* simplex Δ then τG again satisfies (11.6), (11.7) and $\det(\tau G) \neq 0$. Hence we only have to check (11.10) in order for τG

to be another Gram matrix. In that case the corresponding simplex Δ^τ is again rational and Δ^τ is spherical or hyperbolic depending on the sign of $\det(\tau G)$. In the 3-dimensional case notice also that a rational simplex Δ has zero Dehn-invariant so that $[\Delta]$ defines an element in $H_3(\mathrm{Sl}(2,\mathbb{C}))^\pm$. In particular we shall study the following case:

DEFINITION 11.16. An *orthoscheme* $\Delta = \Delta(\alpha,\beta,\gamma)$ is a (spherical or hyperbolic) 3-simplex with 3 perpendicular faces such that the Gram matrix has the form

$$G(\alpha,\beta,\gamma) = \begin{bmatrix} 1 & -a & 0 & 0 \\ -a & 1 & -b & 0 \\ 0 & -b & 1 & -c \\ 0 & 0 & -c & 1 \end{bmatrix}$$

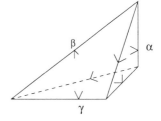

where $a = \cos\alpha$, $b = \cos\beta$, $c = \cos\gamma$, $0 < \alpha, \gamma \le \pi/2$, $0 < \beta < \pi$.

The matrix (11.16) satisfies (11.10) if and only if

$$\cos^2\beta < \sin^2\alpha, \sin^2\gamma, \quad \text{that is}, \quad 0 < \alpha, \gamma < \pi/2 < \alpha+\beta, \beta+\gamma.$$

Also

$$\det G(\alpha,\beta,\gamma) = \sin^2\alpha \cdot \sin^2\gamma - \cos^2\beta.$$

Thus, $G(\alpha,\beta,\gamma)$ is the Gram matrix of a *spherical* orthoscheme if and only if

(11.17,+) $$\cos^2\beta < \sin^2\alpha \cdot \sin^2\gamma$$

whereas it is the Gram matrix of a *hyperbolic* orthoscheme if and only if

(11.17,−) $$\sin^2\alpha \cdot \sin^2\gamma < \cos^2\beta < \sin^2\alpha, \sin^2\gamma.$$

It is easy to see that every simplex can be decomposed into orthoschemes [project one vertex perpendicularly onto the opposite face and its edges].

In particular this reduces the calculation of the volume of a simplex to that of an orthoscheme. Already Lobachevsky and Schläfli calculated the volume of an orthoscheme in terms of certain analytic functions of α, β, γ (cf. [**Coxeter, 1935**]). However notice the following particularly simple case:

EXAMPLE 11.18. The spherical orthoscheme $\Delta(\frac{\pi}{2}, \beta, \frac{\pi}{2}), 0 < \beta < \pi$. Taking the double we clearly get a cone on a triangle with angles $(\frac{\pi}{2}, \frac{\pi}{2}, \beta)$. Hence the union of 4 copies is s.c. to the lune $L(\beta)$, so by the unique divisibility of $\mathcal{P}(S^3), \Delta(\frac{\pi}{2}, \beta, \frac{\pi}{2})$ is s.c. to $L(\beta/4)$. In particular

$$\mathrm{Vol}_{S^3}\left(\Delta\left(\frac{\pi}{2}, \beta. \frac{\pi}{2}\right)\right) = \frac{1}{4}\,\mathrm{Vol}_{S^3}(L(\beta)) = \frac{1}{4} \cdot \frac{\beta}{2\pi} \cdot \mathrm{Vol}_{S^3}(S^3)$$

$$= \frac{1}{4} \cdot \frac{\beta}{2\pi} \cdot 2\pi^2 = \frac{\beta\pi}{4}.$$

Note also that the Dehn-invariant is given by

$$D\left(\Delta\left(\frac{\pi}{2}, \beta, \frac{\pi}{2}\right)\right) = \frac{\pi}{2} \otimes \beta/\pi$$

and that this is zero if and only if β is a rational multiple of π.

In this context let us mention the following question raised by J. Cheeger and J. Simons (1973, cf. [**Cheeger-Simons, 1985**]):

Rational Simplex Problem.   Is the volume of a rational spherical simplex a rational multiple of π^2?

Clearly each of the simplices $\Delta\left(\frac{\pi}{2}, \beta, \frac{\pi}{2}\right)$ in example 11.18 for $\beta \in \mathbb{Q}\pi$ has volume in $\mathbb{Q}\pi^2$ simply because it is s.c. to a rational lune. And so are all other rational simplices for which the volume is known to be a rational multiple of π^2 (cf. [**Dupont-Sah, 1999**, remark following lemma 4.2]). On the other hand *rational orthoschemes* (i.e. orthoschemes $\Delta(\alpha, \beta, \gamma)$ with $\alpha, \beta, \gamma \in \mathbb{Q}\pi$) are very often *not* s.c. to rational lunes as we shall now see (cf. [**Dupont-Sah, 1999**]):

THEOREM 11.19. *Let* $G = G(\alpha, \beta, \gamma)$ *be the Gram matrix of a rational orthoscheme* Δ *of spherical or hyperbolic type.*

a) *For* $\tau \in \mathrm{Gal}(\mathbb{C}/\mathbb{Q})$, *the Gram matrix* $\tau(G)$ *determines again a rational simplex* Δ^τ *of spherical or hyperbolic type.*

b) *In particular for* Δ *spherical it is scissors congruent to a rational lune if and only if*

$$\det(\tau(G)) > 0 \qquad \forall \tau \in \mathrm{Gal}(\mathbb{Q}[e^{i\alpha}, e^{i\beta}, e^{i\gamma}]/\mathbb{Q})$$

c) *Suppose* p, q *are distinct prime numbers with* $1/6 > 4/q + 1/p$. *Then there is a spherical orthoscheme* $\Delta = \Delta(\alpha, \beta, \alpha)$ *with* $e^{i\alpha}$ *and* $e^{i\beta}$ *of order* $2p$ *respectively* $2q$ *and a Galois automorphism* τ *such that* Δ^τ *is hyperbolic. In particular* Δ *is* not *scissors congruent to a lune.*

Proof: a) Notice that the matrix $G = G(\alpha, \beta, \gamma)$ in definition 11.16 has positive definite minors corresponding to the (2,2) and (3,3) entries, and that this property is preserved by the Galois action. It follows that the real symmetric matrix τG cannot have signature (2,2) and therefore determines a spherical or hyperbolic simplex depending on the sign of $\det(\tau G)$.

b) By a) and theorem 11.14 the spherical simplex Δ defines an element $z \in H_3(\mathrm{Sl}(2, \bar{\mathbb{Q}}))^+$ such that

$$\sigma(\tau z) = [\Delta^\tau] \in \mathcal{P}(S^3)/\Sigma(\mathcal{P}(S^2)), \quad \forall \tau \in \mathrm{Gal}(\bar{\mathbb{Q}}/\mathbb{Q}).$$

Hence all the Borel regulators $\{r_2^\tau\}$ vanish on z, so that $z \in \mathbb{Q}/\mathbb{Z} = H_3(\mu_{\bar{\mathbb{Q}}})$, and hence Δ is s.c. to a rational lune by corollary 10.21.

c) First let us note that the second statement follows from the first. In fact, since $\mathrm{Vol}_{\mathcal{H}^3}(\Delta^\tau) \neq 0$ it follows that $[\Delta^\tau] \in \mathcal{P}(\mathcal{H}^3)$ has infinite order and hence the same is true for $[\Delta] \in \mathcal{P}(S^3)/\Sigma(\mathcal{P}(S^2))$.

As for the construction of $\Delta = \Delta(\alpha, \beta, \alpha)$ we just have to satisfy (11.17,+), that is, the inequality

$$\cos^2 \beta < \sin^4 \alpha; \text{ equivalently }, 0 < |\cos \beta| < \sin^2 \alpha.$$

Since $q > 3$ and $p > 5$ we can choose β and α so that $\pi/3 < \beta < 2\pi/3$ and $\pi/4 < \alpha < 3\pi/4$, that is,

$$0 < |\cos \beta| < \frac{1}{2} < \sin^2 \alpha.$$

In order to find τ we observe that since p and q are distinct odd primes, the corresponding cyclotomic fields of $2p$th and $2q$th roots of unity are linearly disjoint. Thus, we can find Galois automorphisms carrying $\exp(i\alpha)$ and $\exp(i\beta)$ independently to arbitrary $2p$th and $2q$th roots of unity $\exp i\alpha'$ and $\exp i\beta'$. That is, we must find α' and β' satisfying the following inequalities:

$$\sin^4 \alpha' < \cos^2 \beta' < \sin^2 \alpha'.$$

We begin by taking $0 < \alpha' < \pi/6$ so that $0 < \sin \alpha' < 1/2$. We then have

$$\sin(\alpha'/2) - \sin^2 \alpha' = \sin(\alpha'/2) \cdot [1 - 2\cos(\alpha'/2) \cdot \sin \alpha'] > 0.$$

In other words it is enough to find α' and β' so as to satisfy

$$0 < \alpha'/2 < |\pi/2 - \beta'| < \alpha' < \pi/6.$$

By our assumption $p/6 - 4p/q > 1$ and thus we can find an integer t so that $4p/q < t < p/6$. We take $\alpha' = t\pi/p$ so that $\alpha' < \pi/6$ holds. Between α' and $\alpha'/2$ we have an angular interval $t\pi/2p > 2\pi/q$. Since the primitive $2q$-th of unity together with -1 are uniformly distributed around the unit circle with successive angular gaps of $2\pi/q$, we may invoke Dirichlet's Box Principle to find β so that $|\pi/2 - \beta|$ falls strictly between $\alpha'/2$ and α'. The strictness is a consequence of the fact that p and q are distinct primes. We note also that β' could not be equal to π so that $\exp i\beta'$ is a primitive $2q$th root of unity. The Galois automorphism τ taking $\exp i\alpha$ and $\exp i\beta$ to $\exp i\alpha'$ and $\exp i\beta'$ is now of the desired form. $\qquad \square$

Remark: There are also other examples of spherical orthoschemes with a hyperbolic Galois conjugate apart from the infinite family constructed in theorem 11.19 c) above. Thus the spherical orthoscheme $\Delta(3\pi/5, \pi/3, 3\pi/5)$ has the hyperbolic conjugate $\Delta(\pi/5, \pi/3, \pi/5)$ which

is the fundamental domain for the hyperbolic Coxeter group corresponding to the diagram (cf. Bourbaki, *Groupes et algèbre de Lie*, chapter
V, p. 133)

$$\bullet \xrightarrow{\ 5\ } \bullet \xrightarrow{\ 3\ } \bullet \xrightarrow{\ 5\ } \bullet$$

As a consequence of the above theorem we get the following relation
between the Non-Euclidean Hilbert's 3rd Problem (cf. chapter 1) and
the Rational Simplex Problem mentioned above:

COROLLARY 11.20. *Suppose that the s.c. classes of 3-dimensional
spherical polyhedra are determined by the volume and Dehn invariant.
Then the volume of a rational spherical orthoscheme* Δ *is an irrational
multiple of* π^2 *if and only if it has a hyperbolic Galois conjugate. In
particular this holds for the infinitely many simplices in theorem 11.19,
c).*

$\square$

Remark: In other words, each of these orthoschemes with a hyperbolic
Galois conjugate provides a negative answer to either the Rational Simplex Problem or the Non-Euclidean Hilbert's 3rd Problem. In any case
let Δ be one of these simplices and let $L = L(\theta)$ be the lune with the
same volume. Then Δ and L are *not* s.c. and thus provides a spherical
analogue to the question asked by Hilbert in the Euclidean case. For the
construction of a similar pair of hyperbolic polyhedra which are *not* s.c.
but have the same volume we refer to [**Dupont-Sah, 1999**, theorem
1.3].

CHAPTER 12

Rigidity of Cheeger-Chern-Simons invariants

In chapter 10 we showed that the s.c. classes of polyhedra defined over the field of algebraic numbers are determined by the volume, the Dehn invariant and the twisted Borel regulators $\{r_2^\alpha \mid \alpha \in \mathrm{Gal}(\bar{\mathbb{Q}}/\mathbb{Q})\}$. Now these are only the *imaginary* parts of the twisted Cheeger-Chern-Simons invariants $\{\hat{C}_2^\alpha\}$ and one could hope that the *real* part would give some further information on the s.c. classes of *all* polyhedra, especially since we are twisting with the larger group $\mathrm{Gal}(\mathbb{C}/\mathbb{Q})$. However in this chapter we shall show that this is not the case. In general, similarly to the definition in (10.17) we consider the tensor product

$$\hat{C}_{k_1\cdots k_l}^{\alpha_1\cdots\alpha_l} = \hat{C}_{k_1}^{\alpha_1} \otimes \cdots \otimes \hat{C}_{k_l}^{\alpha_l} \in H^{2(k_1+\cdots+k_l)-l}(\mathrm{Gl}(n,\mathbb{C}),(\mathbb{C}/\mathbb{Q})^{\otimes l})$$

where $k_1 < \cdots < k_l$ and $\alpha_1,\cdots,\alpha_l \in \mathrm{Gal}(\mathbb{C}/\mathbb{Q})$. Note that we only need to take coefficients in $\mathbb{C}/\mathbb{Q}$ since the torsion is determined by Suslin's theorem 9.22. With this notation we shall show the following.

THEOREM 12.1. *For $k > 1$, the twisted Cheeger-Chern-Simons invariants $\hat{C}_k^\alpha : H_{2k-1}(\mathrm{Gl}(n,\mathbb{C})) \to \mathbb{C}/\mathbb{Q}, \alpha \in \mathrm{Gal}(\mathbb{C}/\mathbb{Q})$, satisfy:*

i) *For all $Z \in H_*(\mathrm{Gl}(n,\mathbb{C}))$ there exists $Z' \in H_*(\mathrm{Gl}(n,\bar{\mathbb{Q}}))$ such that*
$$\hat{C}_{k_1\cdots k_l}^{\alpha_1\cdots\alpha_l}(Z) = \hat{C}_{k_1\cdots k_l}^{\alpha_1\cdots\alpha_l}(Z')$$
for all α_i and all $k_i > 1$.
ii) *If $\alpha,\beta \in \mathrm{Gal}(\mathbb{C}/\mathbb{Q})$ have the same restriction to $\bar{\mathbb{Q}}$ then $\hat{C}_k^\alpha = \hat{C}_k^\beta$.*
iii) *If $Z_1, Z_2 \in H_*(\mathrm{Sl}(n,\mathbb{C}))$ satisfy*
$$r_{k_1\cdots k_l}^{\alpha_1\cdots\alpha_l}(Z_1) = r_{k_1\cdots k_l}^{\alpha_1\cdots\alpha_l}(Z_2)$$
for all twisted Borel regulators $r_{k_1\cdots k_l}^{\alpha_1\cdots\alpha_l}$ (cf. 10.17), then also
$$\hat{C}_{k_1\cdots k_l}^{\alpha_1\cdots\alpha_l}(Z_1) = \hat{C}_{k_1\cdots k_l}^{\alpha_1\cdots\alpha_l}(Z_2)$$

for all α_i *and all* $k_i > 1$.

The proof of this theorem uses some of the same ideas used by [**Reznikov, 1995**] in his proof of the Bloch conjecture (theorem 10.20). The main ingredient is the *rigidity* property for the Cheeger-Chern-Simons classes. For this let us first recall the original construction of these:

Let $E \to M$ be a principal $\text{Gl}(n, \mathbb{C})$-bundle with a flat connection A and let $Z \subseteq M$ a $(2k - 1)$-cycle. Since the odd-dimensional homology of $B \text{Gl}(n, \mathbb{C})$ vanishes one can choose an embedding of bundles

$$
\begin{array}{ccc}
E & \subset & \tilde{E} \\
\downarrow & & \downarrow \\
M & \subset & \tilde{M}
\end{array}
$$

(i.e. E is the restriction of $\tilde{E}$ to M) such that $Z = \partial X$ for some chain $X \subseteq \tilde{M}$. Also choose $\tilde{A}$ a connection on $\tilde{E}$ extending A. Then by definition the evaluation of $\hat{C}_k(A)$ on $Z = \partial X$ is given by

$$(12.2) \qquad \langle \hat{C}_k(A), Z \rangle = \int_X C_k(F_{\tilde{A}})$$

where $F_{\tilde{A}}$ is the curvature form of $\tilde{A}$ and C_k is the kth Chern polynomial. It is straight forward to show that the reduction mod $\mathbb{Z}$ of the expression in (12.2) is independent of the various choices. With this definition it is now easy to prove the following (cf. [**Cheeger-Simons, 1985**]):

THEOREM 12.3 (Rigidity). *Let* $E \to M$ *be a principal* $\text{Gl}(n, \mathbb{C})$-*bundle and let* $\{A_t \mid 0 \leq t \leq 1\}$ *be a smooth family of flat connections. If* $k > 1$ *then*

$$\langle \hat{C}_k(A_1), Z \rangle = \langle \hat{C}_k(A_0), Z \rangle$$

for Z *any* $2k - 1$-*cycle in* M.

Proof: For this put $I = [0, 1]$ and apply (12.2) to $\tilde{M} = M \times I$, $\tilde{E} = E \times I$ and M replaced by $M \times \{0, 1\}$. Then the family $\{A_t\}$ of flat connections on E defines a connection $\tilde{A}$ in $\tilde{E}$ with curvature form

$$F_{\tilde{A}} = d\tilde{A} + \tilde{A} \wedge \tilde{A} = dt \wedge \frac{\partial \tilde{A}}{\partial t}$$

so in particular $F_{\tilde{A}}^k = 0$ for $k > 1$. Taking $X = Z \times I$ in (12.2) we get

$$\langle \hat{C}_k(A_1), Z \rangle - \langle \hat{C}_k(A_0), Z \rangle = \langle \hat{C}_k(\tilde{A}), \partial X \rangle$$
$$= \int_X C_k(F_{\tilde{A}}) = 0$$

$\square$

In the following we put $G = \mathrm{Gl}(n, \mathbb{C})$ and we shall reformulate theorem 12.3 in terms of the homomorphism

$$\hat{C}_k : H_{2k-1}(G) \to \mathbb{C}/\mathbb{Z}$$

where

$$H_*(G) = H(\bar{B}_*(G))$$

is given by the bar complex as in chapter 4. A chain in $\bar{B}_q(G)$ is a finite sum

$$(12.4) \qquad Z = \sum_{\nu=1}^{N} \eta_\nu [g_1^\nu | \dots | g_q^\nu], \quad \eta_\nu \in \mathbb{Z}, \quad g_i^\nu \in G,$$

and Z is a q-cycle if $\partial Z = 0$, that is

$$(12.5) \qquad \sum_{\nu=1}^{N} \sum_{i=0}^{q} (-1)^i \eta_\nu \varepsilon_i [g_1^\nu | \dots | g_q^\nu] = 0$$

where $\varepsilon_i, \quad i = 0, \dots, q$, are the usual boundary operators

$$\varepsilon_i [g_1 | \dots | g_q] = \begin{cases} [g_2 | \dots | g_q], & \text{for } i = 0 \\ [g_1 | \dots | g_i g_{i+1} | \dots g_q], & \text{for } 0 < i < q, \\ [g_1 | \dots | g_{q-1}], & \text{for } i = q. \end{cases}$$

Given a cycle Z the finite set of group elements $\{g_i^\nu\}$ generates a subgroup $\Gamma \subseteq G$ which we consider as an abstract discrete group. Then the expression (12.4) clearly defines a cycle $Z_\Gamma \in \bar{B}_q(\Gamma)$ such that the inclusion $f_0 : \Gamma \hookrightarrow G$ takes Z_Γ to Z. Now f_0 determines a flat connection A on the associated $\mathrm{Gl}(n, \mathbb{C})$-bundle over $B\Gamma$ and clearly

$$(12.6) \qquad \hat{C}_k(Z) = \langle \hat{C}_k(A), Z_\Gamma \rangle.$$

On the other hand f_0 is a point in the *representation variety* defined by

$$R(\Gamma) = \mathrm{Hom}(\Gamma, G)$$

This is well-known to be an affine algebraic variety defined over $\mathbb{Q}$ (cf. [**Raghunatan, 1972**, chapter VI]). An arc $f_t, t \in [0, 1]$, in $R(\Gamma)$ determines similarly a family of flat connections. Hence by theorem 12.3 we get for each $z \in H_{2k-1}(\Gamma), k > 1$, that the function on $R(\Gamma)$ defined by

$$(12.7) \qquad f \mapsto \hat{C}_k(f_*(z))$$

is constant on each arcwise connected component of $R(\Gamma)$. In particular since

$$R(\Gamma) = R(\Gamma)_0 \cup \cdots \cup R(\Gamma)_m$$

is a finite union of *irreducible components* over $\bar{\mathbb{Q}}$ and since each $R(\Gamma)_i$ is arcwise connected by a theorem of Lefschetz (cf. [**Lefschetz, 1953**, p.97] or [**Mumford, 1976**, I, Corollary 4.16]), we conclude the following:

COROLLARY 12.8. *For $z \in H_{2k-1}(\Gamma), k > 1$, the function on $R(\Gamma)$ defined by (12.7) is constant on each irreducible component $R(\Gamma)_i$.* $\square$

Proof of theorem 12.1.   i) Continuing the notation above let us assume $f_0 \in R(\Gamma)_0$. Then by Hilbert's Null-Stellen Satz there is an algebraic point $f' \in R(\Gamma)_{0\mathbb{Q}}$. Consider the associated cycle

$$Z' = f'_*(Z_\Gamma) \in \bar{B}_q(G_{\bar{\mathbb{Q}}})$$

We shall show that if $2(k_1 + \cdots + k_l) - l = q, k_i > 1$, and $\alpha_1, \ldots, \alpha_l \in \mathrm{Gal}(\mathbb{C}/\mathbb{Q})$ then

$$(12.9) \qquad \hat{C}_{k_1 \cdots k_l}^{\alpha_1 \cdots \alpha_l}(Z') = \hat{C}_{k_1 \cdots k_l}^{\alpha_1 \cdots \alpha_l}(Z).$$

The diagonal $\Delta : \Gamma \to \Gamma \times \cdots \times \Gamma$ (l factors) has an induced map in homology with rational coefficients such that

$$\Delta_*([Z_\Gamma]) = \sum z_{i_1} \otimes \cdots \otimes z_{i_l} \qquad z_{i_j} \in H_*(\Gamma, \mathbb{Q}).$$

Hence by definition

$$\hat{C}_{k_1 \cdots k_l}^{\alpha_1 \cdots \alpha_l}(Z) = \sum \hat{C}_{k_1}(\alpha_{1_*} f_{0_*}(z_1)) \otimes \cdots \otimes \hat{C}_{k_l}(\alpha_{i_{l_*}} f_{0_*}(z_{i_l}))$$

and similarly for $\hat{C}_{k_1 \cdots k_l}^{\alpha_1 \cdots \alpha_l}(Z')$ with f_0 replaced by f'. But since $f', f_0 \in R(\Gamma)_0$ also $\alpha_i(f')$ and $\alpha_i(f_0)$ both lie in the same irreducible component of $R(\Gamma)$ so by corollary 12.8

$$\hat{C}_{k_i}(\alpha_{i_*} f_{0_*}(z_i)) = \hat{C}_{k_i}(\alpha_i(f_0)_*(z_i)$$
$$= \hat{C}_{k_i}(\alpha_i(f')_*(z_i)) = \hat{C}_k(\alpha_{i_*} f'_*(z_i))$$

which proves (12.9).

ii) clearly follows from i).

iii) also clearly follows from i) together with Borel's theorem 10.18.

$\square$

CHAPTER 13

Projective configurations and homology of the projective linear group

For the study of the structure of the scissors congruence groups in spherical and hyperbolic 3-space a basic ingredient was the exact sequence in theorem 8.19 relating the homology of $\mathrm{Sl}(2, F)$, F a field of characteristic 0, with the group $\mathcal{P}_F$ generated by cross-ratios of four points on the projective line $P^1(F)$. In the remaining chapters we shall outline a similar program for studying the homology of the projective linear group $\mathrm{PGl}(n + 1, F)$ in terms of configurations of points in projective n-space $P^n(F)$. This goes back several years to unpublished work together with C.-H. Sah (cf. [**Sah, 1989**, problem 4.12]) but it was never completed. In the meantime it has been bypassed by the work of A. Goncharov (cf. [**Goncharov, 1995**]) and others (see e.g. [**Beilinson et. al, 1991**], [**Cathelineau, 1993**], [**Cathelineau, 1995**]).

In the following chapters the field F is assumed algebraically closed for convenience and we put $G = \mathrm{PGl}(n + 1), F)$ which clearly acts on the projective n-space $P^n = P^n(F) = P(F^{n+1})$. For a linear subspace $V \subseteq F^{n+1}$, $P(V) \subset P^n$ denotes the corresponding projective subspace, and $P(\{0\}) = \emptyset$. Also for convenience we consider all chain complexes with $\mathbb{Q}$ coefficients. Furthermore we change the notation slightly and let $C_*(P^n)$ denote the *alternating chain complex* of formal $\mathbb{Q}$-linear combinations of *projective configurations* , i.e. $(q + 1)$-tuples of points $(a_0, \ldots, a_q), a_i \in P^n$, with the identifications for any permutation π:

$$(13.1) \qquad (a_{\pi(0)}, \ldots, a_{\pi(q)}) = \mathrm{sign}\,\pi \quad (a_0, \ldots, a_q).$$

In particular $(a_0, \ldots, a_q) = 0$ if $a_i = a_j$ for some $i \neq j$. The boundary map is given by the usual formula

$$(13.2) \qquad \partial(a_0, \ldots, a_q) = \sum_{i=0}^{q} (-1)^i (a_0, \ldots, \hat{a}_i, \ldots, a_q).$$

Also let $\varepsilon : C_0(P^n) \to \mathbb{Q}$ be the augmentation and let $\tilde{C}_*(P^n)$ denote the augmented complex with $\mathbb{Q}$ in degree -1. Again $C_*(P^n)$ is an acyclic chain complex of $\mathbb{Q}[G]$-modules and we shall study the associated hyperhomology spectral sequences (cf. appendix A). However we shall organize the analysis of this in a convenient way taking into account the geometry of P^n.

In general for N_* any chain complex of (left) $\mathbb{Z}[G]$-modules we consider the *hyperhomology*

$$(13.3) \qquad \boldsymbol{H}_k(G, N_*) = H_k(G \backslash (B_*(G) \otimes N_*))$$

where we are taking the homology of the *total* complex and where $B_*(G)$ is the bar complex as defined in chapter 4. If $N_i = 0$ for $i \neq 0$ then clearly

$$(13.4) \qquad \boldsymbol{H}(G, N_*) = H_*(G, N_0).$$

For a chain complex N_* and $l \in \mathbb{Z}$ let $N_i[l] = N_{i-l}$ with the boundary map multiplied by $(-1)^l$. Then from (13.4) and the exact sequence

$$0 \to \mathbb{Q}[-1] \to \tilde{C}_*(P^n) \to C_*(P^n) \to 0$$

we get a natural isomorphism

$$(13.5) \qquad \boldsymbol{H}_k(G, C_*(P^n)) \cong H_k(G, \mathbb{Q}), \quad \text{for all } k.$$

We next define a filtration of $C_*(P^n)$ by $\mathbb{Q}[G]$ modules

$$(13.6) \qquad C_*(P^n) = \mathcal{F}_{n,*} \supseteq \mathcal{F}_{n-1,*} \supseteq \cdots \supseteq \mathcal{F}_{0*} \supseteq \mathcal{F}_{-1,*} = 0$$

as follows. A configuration $(a_0, \ldots, a_q), a_i \in P^n$, is called *decomposable at level* p if there is a direct decomposition of F^{n+1} into non-zero summands

$$(13.7) \qquad F^{n+1} = V_0 \oplus \cdots \oplus V_{n-p}$$

such that $a_j \in P(V_0) \cup \cdots \cup P(V_{n-p})$ for all $j = 0, \ldots, q$. If $p = 0$ then the $P(V_j)$'s are just single points in which case the configuration either has repetitions i.e. represents 0 in $C'_q(P^n)$, or $q \leq n$ and $(a_0, \ldots, a_q)$ is an *independent* set of points. We now let

$$\mathcal{F}_{p,*} = \mathcal{F}_p C_*(P^n) \subseteq C_*(P^n)$$

be the subcomplex generated by configurations which are decomposable at level p. In particular

(13.8) $$C_*^{\text{indep}}(P^n) = \mathcal{F}_0 C_*(P^n)$$

is generated by the *independent configurations*. We also have the corresponding augmented complexes $\tilde{\mathcal{F}}_{p,*}$, $\tilde{C}_*^{\text{indep}}(P^n)$, with $\mathbb{Q}$ in degree -1. Finally for $V \subseteq F^{n+1}$ a linear subspace we put

(13.9) $$\mathcal{F}_{p,*}(V) = C_*(P(V)) \cap \mathcal{F}_{p,*}$$

and similarly for the augmented complexes.

The filtration $\mathcal{F}_p$ is clearly stable under the action by $\text{PGl}(n+1, F)$ and therefore gives a corresponding filtration on $C_*(P^n) \otimes_{\mathbb{Z}[G]} B_*(G)$. Hence by (13.5) we obtain a spectral sequence

(13.10) $$E_{p,q}^r \Rightarrow H_{p+q}(G, \mathbb{Q})$$

with

(13.11) $$\mathbf{E}_{p,q}^1 = \mathbf{H}_{p+q}(G, \mathcal{F}_{p,*}/\mathcal{F}_{p-1,*}).$$

Now if a configuration is decomposable at level p but *not* at level $p-1$, then, in the decomposition (13.7) above the subspaces V_i of dimension bigger than 1 are unique up to permutation (being spanned by a subset of the configuration). Let us arrange these $V_1, V_2, \ldots, V_k$ according to non-decreasing dimension $2 \leq n_1 + 1 \leq n_2 + 1 \leq \cdots \leq n_k + 1$, and let V_0 denote a complement to $V_1 \oplus V_2 \oplus \cdots \oplus V_k$ containing the 1-dimensional summands. Then our configuration σ has the form

$$\sigma = (a_0, \ldots, a_{q_0}, \ldots, a_{q_0+q_1}, \ldots, a_{q_0+\cdots+q_k})$$

with $(a_0, \ldots, a_{q_0})$ independent in $P(V_0)$ and $(a_{q_i+1}, \ldots, a_{q_{i+1}})$ indecomposable in $P(V_{i+1})$, $i = 0, \ldots, k - 1$. It follows that

(13.12)

$$\mathcal{F}_{p,*}/\mathcal{F}_{p-1,*} \cong \bigoplus_{V_1 \oplus \cdots \oplus V_k \subseteq F^{n+1}} \left[\tilde{C}_*^{\text{indep}}(P(V_0)) \otimes \right.$$

$$\left. \otimes (\mathcal{F}_{n_1,*}(V_1)/\mathcal{F}_{n_1-1,*}) \otimes \cdots \otimes (\mathcal{F}_{n_k,*}(V_k)/\mathcal{F}_{n_k-1,*}) \right][k]$$

where the sum is taken over all direct sums as above and where

$$V_0 = F^{n+1}/(V_1 \oplus \cdots \oplus V_k)$$

has dimension $n_0 + 1 \geq 0$ (so that $\tilde{C}_*^{\text{indep}}(P(V_0)) = \mathbb{Q}[-1]$ for $n_0 = -1$). In particular for the decomposition

$$F^{n-n_0} = F^{n_1+1} \oplus \cdots \oplus F^{n_k+1}$$

the stabilizer for the action by $\text{Gl}(n + 1, F)$ is the semi-direct product of the subgroup of matrices of the form

$$\begin{pmatrix} g_0 & & & 0 \\ \cdot & g_1 & & \\ \cdot & & \ddots & \\ \cdot & 0 & & g_k \end{pmatrix} \qquad g_i \in \text{Gl}(n_i + 1, F), i = 0, \ldots, k,$$

with the group of permutations of the blocks of equal rank. As before, using the "center kills" and Shapiro's lemma (lemmas 5.4 and 5.5) we can now summarize our discussion in the following

THEOREM 13.13. *For* $G = \text{PGl}(n+1, F)$ *there is a spectral sequence* $E_{p,q}^r$ *converging to* $H_*(G, \mathbb{Q})$ *such that* $E_{0,*}^1 = \boldsymbol{H}(G, C_*^{\text{indep}}(P^n))$ *and for* $p > 0$,

$$E_{p,*}^1 \cong \bigoplus \left[\boldsymbol{H}(G(n_0), \tilde{C}_*^{\text{indep}}(P^{n_0})) \otimes \right.$$

$$\left. \otimes \boldsymbol{H}(G(n_1), C_*(P^{n_1})/\mathcal{F}_{n_1-1,*}) \hat{\otimes} \ldots \hat{\otimes} \boldsymbol{H}(G(n_k), C_*(P^{n_k})/\mathcal{F}_{n_k-1,*}) \right][k]$$

where the sum is taken over all $n_0, \ldots, n_k$ *satisfying*

$$-1 \leq n_0, \ 1 \leq n_1 \leq \cdots \leq n_k, \quad n_0 + \cdots + n_k + k = n, \quad n_0 + k = n - p.$$

Here $G(n_i) = \text{PGl}(n_i + 1, F)$, $i = 0, \ldots, k$, and $\hat{\otimes}$ indicates taking coinvariants for the induced action by the group permuting the factors

$G(n_i)$ of equal rank (multiplied by the sign of the corresponding permutation of the standard basis).

By theorem 13.13 we are left with 3 problems:

13.14.i) Compute $\boldsymbol{H}(G, C_*^{\mathrm{indep}}(P^n))$.
13.14.ii) Describe $\boldsymbol{H}(G, C_*(P^n)/\mathcal{F}_{n-1,*})$.
3.14.iii) Determine the differentials in the spectral sequence.

For the remainder of this chapter we shall deal with problem (13.14.i) and defer the other two problems to the next chapters. A direct approach is the usual hyperhomology spectral sequence

$$''E_{**}^r \Rightarrow \boldsymbol{H}(G, C_*^{\mathrm{indep}}(P^n))$$

with

(13.15) $$''E_{p,q}^2 = H_q(H_p(G, C_*^{\mathrm{indep}}(P^n))).$$

Here

(13.16)
$$''E_{p,q}^1 = H_p(G, C_q^{\mathrm{indep}}(P^n))$$
$$= \begin{cases} H_p(G(e_1, \ldots, e_{q+1}) \otimes \mathbb{Q}(e_1, \ldots, e_{q+1}), & q \leq n \\ 0 & q > n \end{cases}$$

where $G(e_1, \ldots, e_{q+1}) \subseteq G = \mathrm{PGl}(n+1, F)$ denotes the stabilizer of the standard independent configuration $(e_1, \ldots, e_{q+1})$.

Again using "center kills" we get for $q \leq n$

$$''E_{*,q}^1 = H(((F^\times)^{q+1} \times \mathrm{Gl}(n - q, F))/F^\times) \bigotimes_{\mathbb{Z}[\mathfrak{S}_{q+1}]} \mathrm{Alt}_{q+1}$$

where $\mathfrak{S}_{q+1}$ is the symmetric group acting by permuting the $(q + 1)$ factors of $F^\times$ and where Alt_{q+1} is the usual sign representation. As before $\mu_F \subseteq F^\times$ is the group of roots of unity and since F is algebraically closed $F^\vee = F^\times/\mu_F$ is a uniquely divisible group, i.e. a rational vector space. Let us summarize

PROPOSITION 13.17. *The spectral sequence*

$$''E_{**}^r \Rightarrow \boldsymbol{H}_*(G, C^{\mathrm{indep}}(P^n))$$

has $''E^1_{*,q} = 0$ *for* $q > n$, *and for* $q \leq n$

$$''E^1_{*,q} = H_*(((F^\times)^{q+1} \times \mathrm{Gl}(n-q,F))/F^\times) \bigotimes_{\mathbb{Z}[\mathfrak{S}_{q+1}]} \mathrm{Alt}_{q+1}$$

$$\cong \begin{cases} \left(\wedge^*((F^\vee)^{q+1}) \bigotimes_{\mathbb{Z}[\mathfrak{S}_{q+1}]} \mathrm{Alt}_{q+1}\right) \otimes H_*(\mathrm{PGl}(n-q,F)), & q < n \\ \wedge^* \left(((F^\vee)^{n+1}/F^\vee)\right) \bigotimes_{\mathbb{Z}[\mathfrak{S}_{n+1}]} \mathrm{Alt}_{n+1}, & q = n. \end{cases}$$

Furthermore $''d^1 : ''E^1_{*,q} \to ''E^1_{*,q-1}$ *is the symmetrization of the map induced by the inclusion*

$$(F^\times)^{q+1} \times \mathrm{Gl}(n-q,F) \to (F^\times)^q \times \mathrm{Gl}(n-q+1,F)$$

Sending

$$(\lambda_0, \ldots, \lambda_q, g) \mapsto \left(\lambda_0, \ldots, \lambda_{q-1}, \begin{pmatrix} \lambda_q & 0 \\ 0 & g \end{pmatrix}\right)$$

Remark: The last statement is straight forward from the formula (13.2) for the differential. Thus to calculate the $''E^2$-term we need some representation theory of the symmetric group.

EXAMPLE 13.18. $''E^2_{p,n} = 0$ for all p. In fact proposition 13.17 gives a diagram

$$
\begin{array}{ccc}
''E^1_{*,n} & \cong & \wedge^*((F^\vee)^{n+1}/F^\vee) \bigotimes_{\mathbb{Z}[\mathfrak{S}_{n+1}]} \mathrm{Alt}_{n+1} \\
{\scriptstyle ''d^1}\Big\downarrow & & \\
''E^1_{*,n-1} & \cong & \wedge^*((F^\vee)^{n+1}/F^\vee) \bigotimes_{\mathbb{Z}[\mathfrak{S}_n]} \mathrm{Alt}_n
\end{array}
$$

in which the symmetric group in the second line acts only on the first n coordinates and $''d^1$ is just the symmetrization of the identity. It follows that $''d^1$ is injective, hence $E^2_{*,n} = 0$.

EXAMPLE 13.19. $''E^2_{n+1,n-1}$. First notice that

$$''E^1_{n+1,n-1} \cong \wedge^{n+1}(F^\vee) \bigotimes_{\mathbb{Z}[\mathfrak{S}_n]} \mathrm{Alt}_n$$

and similarly to example 13.18 it follows that the kernel of

$$''d^1 : ''E^1_{n+1,n-1} \to ''E^1_{n+1,n-2}$$

is contained in

$$(F^\vee)^{\wedge(n-1)} \wedge \wedge^2(F^\vee) \cong S^{n-1}(F^\vee) \otimes \wedge^2(F^\vee)$$

where S^l denotes the l-th symmetric power. It follows using the diagram in example 13.18 that

$$E^2_{n+1,n-1} \subseteq \left[S^{n-1}(F^\vee) \otimes \wedge^2(F^\vee)\right] \Big/ {}''d^1({}''E^1_{n+1,n})$$

is the kernel of the composite map in the diagram

$$S^{n-1}(F^\vee) \otimes \wedge^2(F^\vee) \xrightarrow{\simeq} \wedge^{n+1}((F^\vee)^{n+1}/F^\vee) \otimes_{\mathbb{Z}[\mathfrak{S}_n]} \mathrm{Alt}_n$$

$$\wedge^{n+1}((F^\vee)^{n+1}/F^\vee) \otimes_{\mathbb{Z}[\mathfrak{S}_{n+1}]} \mathrm{Alt}_{n+1}$$

where the vertical map is the natural projection.

On the other hand it is straightforward using the exact sequence

$$0 \to F^\vee \to (F^\vee)^{n+1} \to (F^\vee)^{n+1}/F^\vee \to 0$$

and the fact that

$$\wedge^k(F^\vee)^{n+1} \bigotimes_{\mathbb{Z}[\mathfrak{S}_{n+1}]} \mathrm{Alt}_{n+1} \cong$$

$$\begin{cases} 0, & \text{for } k \leq n-1 \\ S^n(F^\vee) & \text{for } k = n \\ S^{n+1}(F^\vee) \oplus \wedge^2(F^\vee) \otimes S^{n-1}(F^\vee), & \text{for } k = n+1 \end{cases}$$

to show that there is a natural exact sequence

$$0 \to F^\vee \otimes S^n(F^\vee) \to S^{n+1}(F^\vee) \bigoplus \wedge^2(F^\vee) \otimes S^{n-1}(F^\vee) \to$$

$$\to \wedge^{n+1}((F^\vee)^{n+1}/F^\vee) \bigotimes_{\mathbb{Z}[\mathfrak{S}_{n+1}]} \mathrm{Alt}_{n+1} \to 0.$$

Comparing this with the above we find that

$$E^2_{n+1,n-1} \subseteq \mathrm{im}\left[\eta : F^\vee \otimes S^n(F^\vee) \to \wedge^2(F^\vee) \otimes S^{n-1}(F^\vee))\right]$$
$$= T^{(2,n-1)}(F^\vee)$$

where η is the map given by

$$\eta(a_0 \otimes (a_1 \circ \cdots \circ a_n)) = \sum_i (a_0 \wedge a_i) \otimes (a_1 \circ \cdots \circ \hat{a}_i \circ \cdots \circ a_n)$$

and where $T^{(2,n-1)}$ denotes the *Schur functor* associated with the partition $(2, n - 1)$ (cf. [**Akin-Buchsbaum-Weyman, 1982**]). Since $E^2_{*,q} = 0$ for $q \geq n$ we have thus proved the following:

COROLLARY 13.20. *There is a natural projection*

$$p_n : \mathbf{H}_{2n}\left(G, C_*^{\mathrm{indep}}(P^n)\right) \to E^\infty_{n+1,n-1}$$
$$\subseteq T^{(2,n-1)}(F^\vee) \subseteq \wedge^2(F^\vee) \otimes S^{n-1}(F^\vee)$$

where $T^{(2,n-1)}$ is the Shur functor associated with the partition $(2, n-1)$.

There is an alternative inductive approach to the computation of $H(G, C_*^{\mathrm{indep}}(P^n))$ which we shall now describe. For $\{x_{-p}, \ldots, x_{-1}\}$ an independent set of points in P^n let

$$C_q^{\mathrm{indep}}(P^n)_{\{x_{-p},\ldots,x_{-1}\}} \subseteq C_q^{\mathrm{indep}}(P^n)$$

be the subcomplex generated by $(q + 1)$-tuples $(a_0, \ldots, a_q)$ such that $\{x_{-p}, \ldots, x_{-1}, a_0, \ldots, a_q\}$ is an independent set. Also let the associated augmented complex be denoted by $\tilde{C}_*^{\mathrm{indep}}(P^n)_{\{x_{-p},\ldots,x_{-1}\}}$. Notice that there is a natural projection map

(13.21)
$$\pi : C_*^{\mathrm{indep}}(P^n)_{\{x_{-p},\ldots,x_{-1}\}} \to C_*^{\mathrm{indep}}(P(F^{n+1}/\operatorname{span}\{x_{-p}, \ldots, x_{-1}\})).$$

We now have the following.

PROPOSITION 13.22.   i) *There is an exact sequence of G-modules* $(G = \mathrm{PGl}(n + 1, F))$

$$0 \to C_*^{\mathrm{indep}}(P^n) \xrightarrow{b_0} \bigoplus_{\{x_{-1}\}} \tilde{C}_*^{\mathrm{indep}}(P^n)_{\{x_{-1}\}}[1] \xrightarrow{b_1} \cdots$$

$$\cdots \to \bigoplus_{\{x_{-p},\ldots,x_{-1}\}} \tilde{C}_*^{\mathrm{indep}}(P^n)_{\{x_{-p},\ldots,x_{-1}\}}[p] \xrightarrow{b_p} \cdots$$

$$\cdots \to \bigoplus_{\{x_{-n-1},\ldots,x_{-1}\}} \tilde{C}_*^{\mathrm{indep}}(P^n)_{\{x_{-n-1},\ldots,x_{-1}\}}[n + 1] \to 0$$

ii) *The projection π in (13.21) induces an isomorphism*

$$\boldsymbol{H}\left(G,\bigoplus \tilde{C}_*^{\mathrm{indep}}(P^n)_{\{x_{-p},\ldots,x_{-1}\}}\right) \cong$$

$$\begin{cases} \frac{\wedge^*((F^{\vee})^p)}{\mathfrak{S}_p} \otimes \boldsymbol{H}\left(\mathrm{PGl}(n-p+1),\tilde{C}_*^{\mathrm{indep}}(P^{n-p})\right), & p \leqq n \\ \frac{\wedge^*((F^{\vee})^{n+1}/F^{\vee})}{\mathfrak{S}_{n+1}}[-1], & p = n+1. \end{cases}$$

Proof: i) The boundary map b_p is given by

$$b_p(a_0,\ldots,a_q)_{\{x_{-p},\ldots,x_{-1}\}} = \sum_{i=0}^{q}(-1)^i(a_0,\ldots,\hat{a}_i,\ldots,a_q)_{\{a_i,x_{-p},\ldots,x_{-1}\}}$$

for $p = 0,1,2,\ldots$ and it is easy to see that $b \circ b = 0$. Also we have contracting homotopies $s_p, p = 0,1,\ldots,$ defined by $s_0 = 0$ and for $p \geqq 1$,

$$s_p(a_0,\ldots,a_q)_{\{x_{-p},\ldots,x_{-1}\}} = \sum_{i=1}^{p}(x_i,a_0,\ldots,a_q)_{\{x_{-p},\ldots,\hat{x}_{-i},\ldots,x_{-1}\}}$$

It is easily checked that

$$b_{p-1} \circ s_p + s_{p+1} \circ b_p = (p+q+1)\,\mathrm{id}, \quad p = 0,1,\ldots,$$

which shows that the sequence is exact.

ii) As usual this follows using Shapiro's lemma and the "center kills" lemma (lemmas 5.4 and 5.5). $\qquad\square$

Remark: Note that by the exact sequence

$$0 \to \tilde{C}_*^{\mathrm{indep}}(P^n) \to \tilde{C}_*(P^n) \to C_*(P^n)/C_*^{\mathrm{indep}}(P^*) \to 0$$

we get a natural isomorphism

$$\boldsymbol{H}_i(G,\tilde{C}_*^{\mathrm{indep}}(P^n)) \cong \boldsymbol{H}_{i+1}(G,C_*(P^n)/C_*^{\mathrm{indep}}(P^n)).$$

CHAPTER 14

Homology of indecomposable configurations

We next turn to problem (13.14.ii), that is, we want to study the hyper-homology $\boldsymbol{H}(G, C_*(P^n)/\mathcal{F}_{n-1,*})$ for the group $G = \mathrm{PGl}(n+1, F)$ acting on the alternating chain complex of projective configuration modulo all *decomposable configurations*.

Again in this chapter all chain complexes have rational coefficients unless otherwise specified.

LEMMA 14.1. *A configuration is indecomposable if and only if its isotropy subgroup of* $\mathrm{PGl}(n + 1, F)$ *is finite.*

Proof: The isotropy group of a decomposable configuration clearly contains $(F^\times)^2/F^\times \cong F^\times$ which is infinite. Let us show the converse. It is clearly enough to show that if a projective transformation g keeps an indecomposable configuration σ pointwise fixed then $g = \mathrm{id}$. Since σ is indecomposable it must contain an independent set of $n + 1$ points which, by a change of basis, we can assume to be the standard basis $\{e_1, \ldots, e_{n+1}\}$ of F^{n+1}. It follows that g is given by a diagonal matrix and in particular keeps $P^{n-1} = \mathrm{span}\{e_1, \ldots, e_n\}$ invariant. Let σ' be the projection of $\sigma - \{e_{n+1}\}$ on P^{n-1}. Then we have two cases: 1) σ' is indecomposable in P^{n-1}, or 2) σ' is decomposable in P^{n-1}. In case 1) we may assume by induction that g fixes every point of P^{n-1} and since σ is indecomposable it must contain a point in $P^n - (P^{n-1} \cup \{e_{n+1}\})$ which is also fixed. Hence $g = \mathrm{id}$ in this case. In case 2) we have a non-trivial decomposition $F^n = V_1 \oplus \cdots \oplus V_r$, $r > 1$, such that σ' is contained in $\bigcup_{i=1}^r P(V_i)$, and such that $\sigma'_i = P(V_i) \cap \sigma'$ is indecomposable in $P(V_i)$. Again by induction we may assume $g|P(V_i) = \mathrm{id}$, $i = 1, \ldots, r$. Now for each i the original configuration σ must contain some point outside $P(V_i)$ and projecting to a point of σ'_i, since otherwise σ would

135

be contained in $P(V_1 \oplus \cdots \oplus \hat{V}_i \oplus \cdots \oplus V_r \oplus \text{span}\{e_i\}) \cup P(V_i)$. Since g fixes these points it now follows that $g = \text{id}$ on all of P^n. $\square$

It follows that with rational coefficients the hyperhomology spectral sequence for $C_*(P^n)/\mathcal{I}_{n-1,*}$ collapses and we obtain the following:

COROLLARY 14.2. *With rational coefficients there is a natural isomorphism*

$$H(G, C_*(P^n)/\mathcal{I}_{n-1,*}) = H(\bar{C}_*)$$

where $\bar{C}_* = G\backslash C_*(P^n)/\mathcal{I}_{n-1,*}.$ $\square$

Let us put for short

(14.3) $\mathcal{P}_*^n = \mathcal{P}_*^n(F) = H(\bar{C}_*) = H(G\backslash C_*(P^n)/\mathcal{I}_{n-1,*}).$

EXAMPLE 14.4. For $n = 1$ we clearly have (cf. chapter 8)

$$\mathcal{P}_i^1(F) = \begin{cases} 0, & \text{for } i \leq 2, \\ \mathcal{P}_F & \text{for } i = 3. \end{cases}$$

Notice that in degree 3 any configuration is a cycle so that, up to the action by $\text{PGl}(2, F)$ the generators are in standard form given by

$$\{z\} = (e_1, e_2, e_1 + e_2, ze_1 + e_2), \quad z \in F - \{0, 1\}.$$

EXAMPLE 14.5. In $\mathcal{P}_{2n+1}^n$ we have a similar family of cycles: A $2(n + 1)$-*gon* is a configuration $(a_0, \ldots, a_{2n+1})$ such that a_{2i+1} lies on the line through a_{2i} and a_{2i+2}, $i = 0, \ldots, n$, (cf. figure below). Up to the action by $\text{PGl}(n + 1, F)$ it has the standard form

$$\{z\} = (e_1, e_2, e_1 + e_2, e_3, e_1 + e_2 + e_3, \ldots, e_{n+1}, e_1 + e_2 + \cdots + e_{n+1},$$
$$ze_1 + e_2 + \cdots + e_{n+1}) \quad z \in F - \{0, 1\}$$

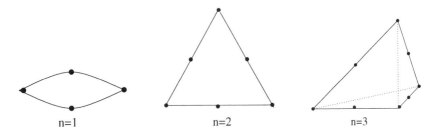

n=1   n=2   n=3

That a $2(n+1)$-gon is a cycle (rationally) follows easily from the following lemma whose proof we leave as a simple exercise in projective geometry.

LEMMA 14.6. *Let $U \subseteq P^n$ be a hyperplane and (a_0, a_1) and (a_0', a_1') two pairs of distinct points outside U. Then there is a projective transformation $g \in \mathrm{PGl}(n+1, F)$ with $ga_0 = a_0', ga_1 = a_1'$ and fixing every point of U, if and only if the lines through (a_0, a_1) respectively (a_0', a_1') intersect U in the same point*   □

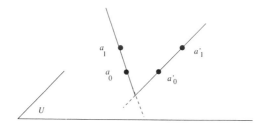

PROPOSITION 14.7. *Let $\bar{C}_* = G\backslash C_*(P^n)/\mathfrak{F}_{n-1,*}$.*

i) *If a configuration σ has all but at most 2 points lying on a hyperplane then $2\sigma = 0$ in $\bar{C}_*$. In particular $\mathcal{P}_i^n = 0$ for $i \leq n+1$.*

ii) *If $\sigma \in \bar{C}_{2n+1}$ is a $2(n+1)$-gon then 2σ is a cycle.*

Proof: i) If σ has at most one point outside a hyperplane then it is decomposable. If $\sigma = (a_0, \ldots, a_q)$ with $a_2, \ldots, a_q \in U$, U a hyperplane and $a_0 \neq a_1$ outside U then by lemma 14.6 we can find $g \in G$ interchanging a_0 and a_1; hence

$$\sigma = (a_0, \ldots, a_q) \sim (a_1, a_0, a_2, \ldots, a_q) = -\sigma.$$

ii) follows since any face of a $2(n+1)$-gon has the form considered in i).   □

Remark: We would have liked to prove that i) $\mathcal{P}_i^n = 0$ for $i \leq 2n$, and ii) $\mathcal{P}_{2n+1}^n$ is generated by $2(n+1)$-gons. However we have only been able to show the first statement for $n \leq 3$ and the second for $n \leq 2$ (cf.

corollary 14.15 below). This is contained in a general result, which we show below (theorem 14.9).

DEFINITION 14.8.   a) A configuration σ in $C_*(P^n)$ is called a k-*wedge*, $k = 1, \ldots, n - 1$, if there are projective subspaces U and V of dimension $n - 1$ respectively k, such that i) $\sigma \subseteq U \cup V$ and ii) $\#\sigma \cap U \cap V = k$.

b) A configuration σ in $C_{2n+1}(P^n)$ is called a *corner* if there are 3 hyperplanes U_1, U_2, U_3 such that i) $\sigma \subseteq U_1 \cup U_2 \cup U_3$, ii) $\#\sigma \cap U_1 \cap U_2 \cap U_3 = n - 2$, and iii) $\#\sigma \cap U_i \cap U_j = n - 1$ for $i \neq j$.

Remark: The following figures are examples of corner configurations for $n = 2$ and 3. Notice that for $n = 2$ a corner with at least 3 points outside a line is necessarily a 6-gon.

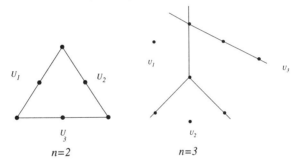

$$n=2 \qquad\qquad n=3$$

Notice that the above configuration for $n = 3$ is *not* a cycle.

We can now state the following general results.

THEOREM 14.9.   *Let* $\bar{C}_* = G\backslash C_*(P^n)/\mathcal{F}_{n-1,*}$ *and* $\mathcal{P}_k^n = H_k(\bar{C}_*)$. *For* $r = 0, 1, 2, \ldots$, *there are explicit chain maps* $f^r : \bar{C}_* \to \bar{C}_*$ *satisfying the following:*

i) $f^0 = \mathrm{id}$ *and* $f^r|\bar{C}_i = f^{r-1}|\bar{C}_i$ *for* $i \leq n + r - 1$.

ii) f^r *is chain homotopic to* $f^{r-1}, r = 1, 2, \ldots$.

iii) *If* $c \in \mathcal{P}_{n+r}^n$, $r \leq n - 1$, *then* $f^r(c)$ *is a linear combination of* r-*wedge configurations.*

iv) *If* $c \in \mathcal{P}_{2n}^n$, $n > 1$, *then* $f^n(c)$ *is also a linear combination of* $(n-1)$-*wedge configurations.*

v) *If* $c \in \mathcal{P}_{2n+1}^n$ *then* $f^{n+1}(c)$ *is a linear combination of* $(n-1)$-*wedge or corner configurations.*

Proof: (Sketch). As a start let us define f^1 and its chain homotopy s^1 to the identity. Let us define $f^1(\sigma)$ for $\sigma = (a_0, \ldots, a_q), q \geq n+1$, a configuration of points $a_i \in P^n$ which we shall assume in general position for convenience. Let $I = (i_0, \ldots, i_{n+1})$ be a sequence of $n+2$ distinct numbers from $\{0, \ldots, q\}$ and put $I' = \{i_0, \ldots, i_{n-1}\}, I'' = \{i_n, i_{n+1}\}$. Let $a^1 = a^1(I)$ be the intersection point between the hyperplane span$\{a_i \mid i \in I'\}$ and the line $\{a_i \mid i \in I''\}$. Furthermore given $j \in I$ put

$$a_\nu^1 = a_\nu^1(I, j) = \begin{cases} a_\nu & \nu \neq j, \quad \nu = 0, \ldots, q, \\ a^1 & \nu = j, \end{cases}$$

With this notation we define

$$t^1(\sigma) = \sum_I (a^1, a_0, \ldots, a_q)$$

and

$$g^1(\sigma) = \sum_I \sum_{j=0}^{n+1} (a_0^1, \ldots, a_q^1).$$

Then

(14.10) $$\partial \circ t^1(\sigma) + t^1 \circ \partial(\sigma) = N_q \sigma - g^1(\sigma)$$

where $N_q > 0$ is the number of sequences I as above. We now define for $\sigma \in \tilde{C}_q$

$$f^1(\sigma) = \frac{1}{N_q} g^1(\sigma) + \left(\frac{1}{N_q} - \frac{1}{N_{q-1}} \right) t^1 \circ \partial(\sigma)$$

and

$$s^1(\sigma) = \frac{1}{N_q} t^1(\sigma).$$

From the identity (14.10) we now conclude

$$\partial \circ s^1 + s^1 \circ \partial = \mathrm{id} - f^1$$

and hence also

$$\partial - \partial \circ f^1 = \partial \circ s^1 \circ \partial = \partial - f^1 \circ \partial.$$

It follows that f^1 is a chain map chain homotopic to the identity and furthermore iii) for $r = 1$ clearly follows from the above construction.

For general $r \geq 1$ we similarly first construct maps g^r and t^r, $r = 1, 2, \ldots$, such that for $\sigma \in \bar{C}_q$

$$(14.11) \qquad \partial \circ t^r(\sigma) + t^r \circ \partial(\sigma) = N_q^r g^{r-1}(\sigma) - g^r(\sigma), \qquad r = 1, 2, \ldots.$$

for suitable positive integers N_q^r. Then the chain maps f^r and chain homotopies $s^r, r = 1, 2, \ldots$, are defined successively using (14.11) such that

$$(14.12) \qquad f^r(\sigma) - \frac{1}{N_q^r} g^r(\sigma) = \text{ some function of } \partial\sigma.$$

For the definition of $g^r(\sigma)$ for $\sigma = (a_0, \ldots, a_q), q \geq n + r$, we again assume $(a_0, \ldots, a_q)$ in general position for convenience. Again let $I = (i_0, \ldots, i_{n+r})$ be a sequence of $n+r+1$ distinct numbers from $\{0, \ldots, q\}$ and let $J = (j_1, \ldots, j_r)$ a subsequence of I satisfying

$$(14.13,\text{i}) \qquad j_k \in \{i_0, \ldots, i_{n+k}\}, \quad k = 1, \ldots, r.$$

Furthermore suppose we have a *partition* $I = I' \cup I''$ satisfying

$$(14.13,\text{ii}) \quad i_0, \ldots, i_{n-1} \in I', \quad i_n, \ldots, i_{n+k} \in I'', \quad \text{if } k \leq \min(n-1, r),$$

$$(14.13,\text{iii}) \qquad i_{2n+s-1} \in I' \text{ if an only if } j_s \in I', \quad s = 1, \ldots, r - n.$$

(In our case (14.13,iii) is only used for $s = 1, 2$.) With this notation we define inductively for $r = 1, 2, \ldots$

$$
\begin{aligned}
a_\nu^r &= a_\nu^r(I', I'', J) \\
&= a_\nu^{r-1}(I' - \{i_r\}, I'' - \{i_r\}, J - \{j_r\}), \qquad \text{for } \nu \neq j_r, \, \nu = 0, \ldots, q,
\end{aligned}
$$

and

$$a_{j_r}^r = a_{j_r}^r(I', I'', J) =$$
$$\text{span}\left\{a_i \mid i \in I', i \neq j_1, \ldots, j_{r-1}\right\} \cap \text{span}\left\{a_i \mid i \in I'', i \neq j_1, \ldots, j_{r-1}\right\}.$$

Notice for the definition of $a_{j_r}^r$ that $\{a_i \mid i \in I, j \neq j_1, \ldots, j_{r-1}\}$ consists of $n+2$ points so that the intersection point exists and is unique. Finally g^r is defined by

$$(14.14) \qquad g^r(\sigma) = \sum_{(I', I'', J)} (a_0^r, \ldots, a_q^r).$$

The maps t^r are defined in a similar way. We shall omit the details as well as the checking of formula (14.11). For the proof of the statements iii)-v) in the theorem we first observe that by (14.12) the maps f^r and g^r have the same image when restricted to the set of cycles in $\bar{C}_*$. Hence by (14.14) it suffices to check that the configurations $\sigma^r = (a_0^r, \ldots, a_q^r)$ have the desired properties. Thus for iii) we just observe that the subspaces

$$U' = \mathrm{span}\{a_{i_0}, \ldots a_{i_{n-1}}\}, \quad U'' = \mathrm{span}\{a_{i_n}, \ldots, a_{i_{n+r}}\}$$

have dimensions $n-1$ respectively r and satisfies

    i) $a_\nu^r \in U' \cup U''$ for all $\nu = 0, \ldots, n+r$,
    ii) $a_{j_1}^r, \ldots, a_{j_r}^r \in U' \cap U''$,

so that σ^r is an r-wedge as claimed.

For the statement iv) we have two cases: 1) $j_1 \in I'$, 2) $j_1 \in I''$. In the case 1) $j_1 \in I'$, we put

$$U' = \mathrm{span}\left(\{a_{i_0}, \ldots, a_{i_{n-1}}, a_{2n}\} - \{a_{j_1}\}\right), \quad U'' = \mathrm{span}\{a_{i_n}, \ldots a_{i_{2n-1}}\}$$

and observe that

    i) $a_\nu^n \in U' \cup U''$ for all $\nu = 0, \ldots, 2n$,
    ii) $a_{j_2}^n, \ldots, a_{j_n}^n \in U' \cap U''$

so that again σ is an $(n-1)$-wedge in this case. In case 2) $j_1 \in I''$, we similarly get an $(n-1)$-wedge lying on $U' \cup U''$ with

$$U' = \mathrm{span}\{a_{i_0}, \ldots, a_{i_{n-1}}\}, \quad U'' = \mathrm{span}\left(\{a_{i_n}, \ldots, a_{i_{2n}}\} - \{a_{j_1}\}\right).$$

Finally for statement v) we have 4 cases:

    1a) $j_1, j_2 \in I'$,   1b) $j_1 \in I', j_2 \in I''$,

    2a) $j_1 \in I'', j_2 \in I'$,   2b) $j_1, j_2 \in I''$

In the cases 1a) and 2b) the configuration σ^{n+1} is again an $(n-1)$-wedge, whereas in the cases 1b) and 2a) it is a corner. Let us just show this in the case 1b) $j_1 \in I'$, $j_2 \in I''$. Then we put

$$U' = \mathrm{span}\left(\{a_{i_0}, \ldots a_{i_{n-1}}, a_{i_{2n}}\} - \{a_{j_1}\}\right)$$
$$U_1'' = \mathrm{span}\{a_{i_n}, \ldots, a_{i_{2n-1}}\},$$
$$U_2'' = \mathrm{span}\left(\{a_{i_n}, \ldots, a_{i_{2n-1}}, a_{i_{2n+1}}\} - \{a_{j_2}\}\right).$$

Then the configuration σ^{n+1} satisfies

i) $a_\nu^{n+1} \in U' \cup U_1' \cup U_2''$, $\quad \nu = 0, \ldots, 2n+1$,

ii) $a_{j_3}^{n+1}, \ldots, a_{j_{n+1}}^{n+1} \in U' \cap U_1' \cap U_2''$.

iii) Also $a_{j_2}^{n+1} \in U' \cap U_1''$, $\quad a_{j_{n+1}}^{n+1} \in U' \cap U_2''$, and $a_l^{n+1} = a_l \in U_1' \cap U''$ for some $l \in \{i_n, \ldots, i_{2n-1}, i_{2n+1}\} - \{j_2, \ldots, j_{n+1}\}$.

This shows that σ^{n+1} is a corner configuration as claimed. $\qquad \square$

COROLLARY 14.15.   i) $\mathcal{P}_q^n = 0$ for $q \leq 2n$ if $n \leq 3$ and for $q \leq n+3$ if $n \geq 3$.

ii) $\mathcal{P}_3^1$ and $\mathcal{P}_5^2$ are generated by the configurations $\{z\}, z \in F - \{0, 1\}$, defined in example 14.5.

iii) $\mathcal{P}_7^3$ is generated by the cycles

$$\{f^4(\sigma) \mid \sigma = (a_0, \ldots, a_7) \text{ any configuration }\}$$

where f^4 is the chain map in theorem 14.9. In particular every cycle is a linear combination of 2-wedge or corner configurations.

Proof: i) If $\sigma \in C_{n+r}(P^n)$ is an r-wedge with $r \leq \min(3, n-1)$ it is easy to see that σ has at most 2 points outside a hyperplane hence represents zero on $\bar{C}_{n+r}$ by proposition 14.7,i). Similarly for $\sigma \in C_{2n}(P^n)$ an $(n-1)$-wedge for $n \leq 3$. Thus i) follows from theorem 14.9 iii) and iv). ii) Since, as noted in the remark following definition 14.8, a corner configuration in P^2 is either a 1-wedge or a 6-gon, each term in $f^3(\sigma)$ is a cycle by proposition 14.7,ii). Hence the statement for $\mathcal{P}_5^2$ follows from theorem 14.9,v). Similarly for P^3 the chain map f^4 is zero in degrees ≤ 6 and hence maps to the cycles in degree 7. Statement iii) is now obvious from theorem 14.9,v). $\qquad \square$

Remark: As in the case of scissors congruence groups $\mathcal{P}_*^n$ can also be expressed in terms of the homology of a Tits complex $\mathcal{T}(P^n)$. In this case a *subspace* is just a union $U = P(V_1) \cup \cdots \cup P(V_k) \subseteq P^n$ such that $V_1 \oplus \cdots \oplus V_k \subseteq F^{n+1}$ is a direct sum of proper linear subspaces (that is $V_i \neq 0, F^{n+1}$), and a *flag* is just a decreasing sequence of such subspaces ordered by inclusion. Then by the usual bicomplex argument we get a natural isomorphism

$$(14.16) \qquad \mathcal{P}_k^n \cong \boldsymbol{H}_{k-1}\left(G, \tilde{C}_*(\mathcal{T}(P^n))\right)$$

where $\tilde{C}_k(\mathcal{T}(P^n))$ denotes the augmented chain complex for $\mathcal{T}(P^n)$. We shall however not pursue this point of view.

CHAPTER 15

The case of $\mathrm{PGl}(3, F)$

As for the problem (13.14.iii), i.e. the problem of determining the differentials in the spectral sequence, we shall restrict to the cases $n = 1, 2$. In particular for the case $n = 2$ we shall relate the homology of $\mathrm{PGl}(3, F)$ with the homology groups $\mathcal{P}_*^2$ of configurations in the projective plane. Again all chain complexes and all homology groups have rational coefficients.

For $n = 1$ we just recapture the results of chapter 8. In fact, for the action of $G = \mathrm{PGl}(2, F)$ on the projective line P^1 the filtration $\mathcal{F}_p C_*(P^1)$ has only two terms

$$C_*^{\mathrm{indep}}(P^1) = \mathcal{F}_0 C_*(P^1) \subseteq \mathcal{F}_1 C_*(P^1) = C_*(P^1)$$

and by corollary 14.2,

$$\boldsymbol{H}_i(G, \mathcal{F}_1/\mathcal{F}_0) = \begin{cases} 0, & i = 0, 1, 2, \\ \mathcal{P}_3^1 = \mathcal{P}_F, & i = 3. \end{cases}$$

Furthermore by proposition 13.22 we have

$$H_i(G, \mathcal{F}_0) \cong \wedge^i \left((F^\vee)^2 / F^\vee \right)^{\mathfrak{S}_2} \cong \begin{cases} \wedge^i(F^\vee) & i \text{ even} \\ 0 & i \text{ odd} . \end{cases}$$

The spectral sequence then reduces to the exact sequence for the pair $(\mathcal{F}_1, \mathcal{F}_0)$, i.e. for $k = 1, 2, \ldots$ we have the exact sequences

(15.1)
$$0 \to H_{2k+1}(\mathrm{PGl}(2, F)) \to \mathcal{P}_{2k+1}^1 \xrightarrow{d^1} \wedge^{2k}(F^\vee) \to H_{2k}(\mathrm{PGl}(2, F)) \to$$
$$\to \mathcal{P}_{2k}^1 \to 0.$$

Since $\mathcal{P}_2^1 = 0$ this agrees with the exact sequence in theorem 8.19, and in particular the differential d^1 on $\mathcal{P}_3^1$ is given by

$$(15.2) \qquad d^1(\{z\}) = 2z \wedge (1 - z), \quad z \in F - \{0, 1\}.$$

We next turn to $n = 2$, i.e. in the following $G = \mathrm{PGl}(3, F)$ acts on the projective plane P^2. Also let us write $C_*^{\mathrm{indep}} = C_*^{\mathrm{indep}}(P^2)$. We then have the following result relating $H_*(G)$ and the groups $\mathcal{P}_*^*$ in low dimensions:

THEOREM 15.3. *Let* $G = \mathrm{PGl}(3, F)$.

i) *The row and column in the diagram below are exact sequences*

$$
\begin{array}{c}
S^2(\wedge^2 F^\vee) \\
\downarrow \\
\mathbf{H}_5(G, C_*^{\mathrm{indep}}) \longrightarrow H_5(G) \longrightarrow \mathcal{P}_5^2 \xrightarrow{d^2} \mathbf{H}_4(G, C_*^{\mathrm{indep}}) \longrightarrow H_4(G) \longrightarrow 0 \\
\searrow{\scriptstyle \lambda^2} \qquad \downarrow \\
\mathcal{P}_4^1 \oplus F^\vee \otimes \mathcal{P}_3^1 \\
\downarrow{\scriptstyle b_*} \\
\wedge^3(F^\vee) \\
\downarrow \\
H_3(G) \\
\downarrow \\
H_3(\mathrm{PSl}(2, F)) \\
\downarrow \\
0
\end{array}
$$

ii) $H_2(G) \cong H_2(\mathrm{PSl}(2, F))$.
iii) λ^2 *is given by*

$$\lambda^2\{z\} = (0, 6z \otimes \{z\}), \quad z \in F - \{0, 1\}$$

iv) $b_* \mid \mathcal{P}_4^1$ *agrees with* d^1 *in (15.1) and* $b_* \mid F^\vee \otimes \mathcal{P}_3^1$ *is given by*

$$b_*(w \wedge \{z\}) = w \wedge z \wedge (1 - z), \quad z, w \in F - \{0, 1\}.$$

Proof: i) The filtration $\mathcal{F}_p C_k(P^2)$ has 3 terms

$$C_*^{\mathrm{indep}}(P^2) = \mathcal{F}_0 \subseteq \mathcal{F}_1 \subseteq \mathcal{F}_2 = C_*(P^2)$$

But in the spectral sequence in theorem 13.13, $E_{1,*}^1$ contains the factor

$$\boldsymbol{H}(G(0), \tilde{C}_*^{\mathrm{indep}}(P^0)) = 0$$

so that $E_{1,*}^1 = 0$. It follows that the spectral sequence reduces to the exact sequence for the pair $(\mathcal{F}_2, \mathcal{F}_1)$ such that

(15.4) $$\boldsymbol{H}(G, \mathcal{F}_1) \xleftarrow{\cong} \boldsymbol{H}(G, \mathcal{F}_0) = \boldsymbol{H}(G, C_*^{\mathrm{indep}}).$$

This clearly gives the exact sequence

(15.5) $$\cdots \to \boldsymbol{H}_i(G, C_*^{\mathrm{indep}}) \to H_i(G) \to \mathcal{P}_i^2 \xrightarrow{d^2} \boldsymbol{H}_{i-1}(G, C_*^{\mathrm{indep}}) \to \cdots$$

For the computation of $\boldsymbol{H}(G, C_*^{\mathrm{indep}})$ we use proposition 13.22 and note that

(15.6)
$$\boldsymbol{H}\left(G, \oplus \tilde{C}_*^{\mathrm{indep}}(P^2)_{\{x_{-1}\}}\right) \cong \wedge^* (F^\vee) \otimes \boldsymbol{H}(\mathrm{PGl}(2, F), \tilde{C}_*^{\mathrm{indep}}(P^1))$$

$$\cong \wedge^* (F^\vee) \otimes \mathcal{P}_*^1[-1]$$

by the last remark in chapter 13. Also note that

$$\boldsymbol{H}\left(G, \oplus \tilde{C}_*^{\mathrm{indep}}(P^2)_{\{x_{-2}, x_{-1}\}}\right) \cong \frac{\wedge^*((F^\vee)^2)}{\mathfrak{S}_2} \otimes \boldsymbol{H}(\mathrm{PGl}(1), \tilde{C}_*^{\mathrm{indep}}(P^0))$$

$$\cong 0.$$

Finally a straightforward calculation gives
(15.7)

$$\boldsymbol{H}_i(G, \oplus \mathbb{Q}_{\{x_{-3}, x_{-2}, x_{-1}\}}) \cong \frac{\wedge^i((F^\vee)^3/F^\vee)}{\mathfrak{S}_3} \cong \begin{cases} \mathbb{Q} & i = 0, \\ 0 & i = 1, \\ \wedge^i(F^\vee), & i = 2, 3, \\ S^2\left(\wedge^2(F^\vee)\right), & i = 4, \end{cases}$$

where S^2 denotes the symmetric square.

It follows that the exact sequence in proposition 13.22, i) for $n = 2$ gives rise to the following exact sequence:

(15.8) $$\to \boldsymbol{H}_i(G, C_*^{\mathrm{indep}}) \to \left[\wedge^*(F^\vee) \otimes \mathcal{P}_*^1\right]_i \to \frac{\wedge^{i-1}((F^\vee)^3/F^\vee)}{\mathfrak{S}_3} \to$$

Since $\mathcal{P}_i^2 = 0$ for $i \leq 4$ we obtain the horizontal exact sequence in theorem 15.3 from the sequence (15.5) which also gives

$$(15.9) \qquad \mathbf{H}_i(G, C_*^{\text{indep}}) \underset{\cong}{\rightarrow} H_i(G) \quad \text{for } i \leq 3.$$

The exact sequence (15.8) in low dimensions together with (15.7) clearly reduces to the vertical exact sequence in theorem 15.3 except for $i \leq 3$ where we obtain the sequence

$$\longrightarrow \wedge^3(F^\vee) \longrightarrow \mathbf{H}_3(G, C_*^{\text{indep}}) \longrightarrow \mathcal{P}_3^1 \xrightarrow{b_*} \wedge^2(F^\vee) \longrightarrow \mathbf{H}_2(G, C_*^{\text{indep}}) \longrightarrow 0.$$

$$\Big\downarrow \cong \qquad\qquad\qquad\qquad\qquad\qquad\qquad \Big\downarrow \cong$$

$$H_3(G) \qquad\qquad\qquad\qquad\qquad\qquad\qquad H_2(G)$$

Comparing this with the sequence (15.1) we get the lower part of the vertical sequence in theorem 15.3 i) as well as the isomorphism in theorem 15.3 ii).

The identification of the maps λ^2 and b_* is straight forward from the definition. $\qquad\qquad\qquad\qquad\qquad\qquad\qquad\qquad\qquad\qquad\qquad\qquad\square$

Remark 1: The results on H_2 and H_3 are special cases of the stability results of [**Suslin, 1984**] (cf. also [**Sah, 1986**],[**Sah, 1989**]).

Remark 2: In the diagram in theorem 15.3 the composed map

$$S^2(\wedge^2(F^\vee)) \to H_4(G)$$

is induced by the natural inclusion $(F^\times)^4/F^\times \to G$. In general let H_* (PGl $(n + 1$, F))^{dec} denote the image of the map induced by the inclusion $(F^\times)^{n+1}/F^\times \to \text{PGl}(n+1, F)$. Then we extract from theorem 15.3 together with (15.1) the following exact sequence

$$(15.10) \qquad \begin{aligned} 0 &\to \frac{H_4(\text{PSl}(2, F))}{H_4(\text{PSl}(2, F))^{\text{dec}}} \to \frac{H_4(\text{PSl}(3, F))}{H_4(\text{PSl}(3, F))^{\text{dec}}} \to \\ &\to \frac{F^\vee \otimes \mathcal{P}_3^1}{\{z \otimes \{z\} \mid z \in F - \{0, 1\}\}} \xrightarrow{b_*} \wedge^3(F^\vee). \end{aligned}$$

Similarly let

$$A = \ker(\mathbf{H}_4(G, C_*^{\text{indep}}) \to H_4(G)) \cap$$
$$\cap \text{im}(S^2(\wedge^2(F^4)) \to \mathbf{H}_4(G, C_*^{\text{indep}})).$$

Then we extract from theorem 15.3 another exact sequence

(15.11)

$$\boldsymbol{H}_5(G, C^{\text{indep}}) \to H_5(\text{PGl}(3, F)) \to \ker(\mathcal{P}_5^2 \xrightarrow{\lambda^2} F^\vee \otimes \mathcal{P}_3^1) \to A \to 0.$$

It is instructive to compare the exact sequences (15.10) and (15.11) with the hyperhomology spectral sequence for $C_*(P^2)$.

Let us end with a few remarks on the homology calculations in the general case of $G = \text{PGl}(n + 1, F)$. Our discussion above for $n = 2$ suggests that we start with the exact sequence in proposition 13.22. In fact, if we replace the 0-th term in the exact sequence by $\tilde{C}_*^{\text{indep}}(P^n)$ we get a resolution of the complex $\mathbb{Q}[-1]$, and together with the last remark of chapter 13 we obtain a hyperhomology spectral sequence in the 4th quadrant of the following form (actually rather in the 2nd quadrant for the transposed indices):

(15.12) $E_{l,-p}^r \Rightarrow H_{l-p}(\text{PGl}(n + 1, F)), \quad 0 \le p \le n + 1, \, p \le l,$

where

(15.13)

$E_{*,-p}^1 =$

$$\begin{cases} \dfrac{\wedge^*((F^\vee)^p)}{\mathfrak{S}_p} \otimes \boldsymbol{H}(\text{PGl}(n - p + 1, F), C_*(P^{n-p})/C_*^{\text{indep}})[p] & p \le n, \\[2mm] \dfrac{\wedge^*((F^\vee)^{n+1}/F^\vee)}{\mathfrak{S}_{n+1}}[n + 1], & p = n + 1. \end{cases}$$

In 15.13 the complex $C_*(P^{n-p})/C_*^{\text{indep}}$ is filtered by the filtration $\mathcal{F}_k$, $k = 1, 2, \ldots, n - p$, defined in chapter 13 and hence the corresponding spectral sequence eventually reduces the calculations to the groups $\mathcal{P}_*^m$ studied in chapter 14. Notice that the vanishing results in corollary 14.15 shows that in (15.13) we have $E_{l,-p}^1 = 0$ at least for $l \le n$. Hopefully $E_{l,-p}^1 = 0$ for $l + p \le 2n$, but this we can only prove for $n \le 3$. In any case the line $l + p = 2n + 1$ is particularly interesting since $E_{2n-p+1,-p}^1$ involves $\mathcal{P}_{2(n-p)+1}^{n-p}$ where the $2(n-p+1)$-gons from example 14.5 occur.

Spectral sequences and bicomplexes

We collect a few facts about spectral sequences, which we are using. As a general reference see [**MacLane, 1963**, chapter XI].

Let R be a commutative ring with unit and $(C_*, \partial)(* \geq 0)$ a chain complex of R-modules. Suppose we have given a *filtration* $F_p = F_p C_*$ of chain complexes

$$0 \subseteq F_0 C_* \subseteq F_1 C_* \subseteq \cdots \subseteq F_p C_* \subseteq \cdots \subseteq C_*.$$

We shall always assume that this filtration is finite in each degree. Then there is an associated *spectral sequence* $\{E^r_{p,q}\}, r = 0, 1, 2, \ldots, \infty$, that is, a sequence of bigraded chain complexes with differentials

(A.1) $$d^r : E^r_{p,q} \to E^r_{p-r, q+r-1}$$

such that

$$E^{r+1}_{p,q} = \ker d^r / d^r (E^r_{p+r, q-r+1})$$

Here $p, q \geq 0$, and if for each k there exists a t such that $F_t C_i = C_i$ for $i \leqslant k$ then $d^r_{p,q} = 0$ for $r > t$ and $p + q \leqslant k$, so that

(A.2) $$E^{t+1}_{p,q} = E^{t+2}_{p,q} = \cdots = E^\infty_{p,q}$$

In fact

(A.3) $$E^0_{p,q} = F_p C_{p+q} / F_{p-1} C_{p+q}$$

and d^r is induced by ∂ restricted to the submodule consisting of elements z with $\partial z \in F_{p-r}$. In particular

(A.4) $$E^1_{p,q} = H_{p+q}(F_p / F_{p-1})$$

and $d^1 : E^1_{p,q} \to E^1_{p-1,q}$ is the boundary map in the exact sequence

$$0 \to F_{p-1} / F_{p-2} \to F_p / F_{p-2} \to F_p / F_{p-1} \to 0$$

Furthermore E^r *converges* to $H(C_*)$ denoted $E^r_{p,q} \Rightarrow H(C_*)$, that is, there is an induced filtration

(A.5) $$F_p(H(C_*)) = \mathrm{im}(H(F_{p,*}) \to H(C_*))$$

such that

$$0 \subseteq F_0 H_* \subseteq \cdots \subseteq F_p H_* \subseteq \cdots \subseteq H(C_*)$$

and there is a canonical isomorphism

(A.6) $$E^\infty_{p,q} \cong F_p H_{p+q} / F_{p-1} H_{p+q}$$

A particular case is the spectral sequence for a *bicomplex* ("double complex"). Thus let $C_{p,q}, p, q \geq 0$, be a bigraded R-module with boundary maps

(A.7) $$\partial' : C_{p,q} \to C_{p-1,q} \quad \partial'' : C_{p,q} \to C_{p,q-1}$$

such that

(A.8) $$\partial' \partial' = 0, \quad \partial'' \partial'' = 0, \quad \partial' \partial'' + \partial'' \partial' = 0.$$

Then the *total complex* C_* with

(A.9) $$C_n = \bigoplus_{p+q=n} C_{p,q} \quad \text{and} \quad \partial = \partial' + \partial''$$

is a chain complex with two filtrations

(A.10) $${}^{\prime}F_p C_n = \bigoplus_{k \leq p} C_{k,n-k} \quad {}^{\prime\prime}F_q C_n = \bigoplus_{k \leq q} C_{n-k,k}$$

with associated spectral sequences $\{{}^{\prime}E^r_{p,q}, {}^{\prime}d^r\}$ and $\{{}^{\prime\prime}E^r_{p,q}, {}^{\prime\prime}d^r\}$ both converging to $H(C_*)$. Here

(A.11) $${}^{\prime}E^1_{p,q} = H_q(C_{p,*}, \partial''), \quad {}^{\prime\prime}E^1_{q,p} = H_p(C_{*,q}, \partial')$$

and

(A.12) $${}^{\prime}E^2_{p,q} = H_p(H_q(C_{*,*}, \partial''), \partial'), \quad {}^{\prime\prime}E^2_{q,p} = H_q(H_p(C_{*,*}, \partial'), \partial'').$$

Remark: In (A.11) and (A.12) we have transposed the indices of the "second" spectral sequence in order to have the notation agreeing with the spectral sequence above for a filtration. In applications however, e.g. for the "hyperhomology" spectral sequence (cf. (A. 26-28) below) we shall keep the indices of the bicomplex in the notation for the spectral sequence and just remember to change the direction of the differentials and the filtration accordingly.

In applications very often one of the spectral sequences collapses from the E^2-term (i.e. $d^r = 0$ for $r > 2$) so that this determines $H(C_*)$. For example

PROPOSITION A.13. *Suppose that for each $q = 0, 1, 2, \ldots$, we have $H_p(C_{*,q}) = 0$ for $p > 0$. Let $C_{-1,q} = \mathrm{coker}[\partial' : C_{1,q} \to C_{0,q}]$. Then the "edge map" e induced by the projection $C_n \to C_{0,n} \to C_{-1,n}$ is a homology isomorphism*

$$e_* : H(C_*) \xrightarrow{\cong} H(C_{-1,*}, \partial'')$$

Proof: By (A.12)

$$``E^2_{q,p} = \begin{cases} 0 & \text{if } p > 0 \\ H_q(C_{-1,*}) & \text{if } p = 0 \end{cases}$$

and so $``E^2_{q,p} = ``E^r_{q,p}$ for $r \geq 2$. Hence

$$H_n(C_*) \cong E^\infty_{n,0} \cong H_n(C_{-1,*})$$

and it is straight forward to check that the isomorphism is induced by e above.   □

Very often bicomplexes are used in connection with the "comparison theorem" to show that certain maps are homology isomorphisms. First let us state it for filtrations:

PROPOSITION A.14. *Let $C_*, \bar{C}_*$ be two chain complexes with filtrations F of C_* and $\bar{F}$ of $\bar{C}_*$ as above (i.e. finite in each degree). Let $f : C_* \to \bar{C}_*$ be a chain map such that $fF_p \subseteq \bar{F}_p, p = 0, 1, 2, \ldots$. Suppose further that for some $t \geq 1$ the induced map of spectral sequences $f^t : E^t_{p,q} \to \bar{E}^t_{p,q}$ is an isomorphism for all p, q. Then also $f_* : H(C_*) \to H(\bar{C}_*)$ is an isomorphism.*

Proof: It follows by induction that all $f^s : E^s_{p,q} \to \bar{E}^s_{p,q}, s \geq t$, are isomorphisms. Hence also $f^\infty : E^\infty_{p,q} \to \bar{E}^\infty_{p,q}$ is an isomorphism. Therefore by (A. 6) we have isomorphisms

$$f_* : F_p H(C_*)/F_{p-1} H(C_*) \xrightarrow{\cong} \bar{F}_p H(\bar{C}_*)/\bar{F}_{p-1} H(\bar{C}_*) \text{ for all } p = 0, 1, 2, \ldots$$

Now by induction on p and the "five-lemma" $f_* : F_p H(C_*) \to \bar{F}_p H(\bar{C}_*)$ are isomorphisms for all p. Hence the proposition follows since the filtration is finite in each degree. □

COROLLARY A.15. *Let* $\{C_{p,q}\}, \{\bar{C}_{p,q}\}, p, q \geq 0$ *be two bicomplexes and* $f : C_{*,*} \to \bar{C}_{*,*}$ *a map of bicomplexes, i.e.* f *commutes with* ∂' *and* ∂''. *Suppose that for some* $t \geq 1$ *the induced map of spectral sequences* $f^t : {}^{`}E^t_{p,q} \to {}^{`}\bar{E}^t_{p,q}$ *is an isomorphism. Then also* $f_* : H(C_*) \to H(\bar{C}_*)$ *is an isomorphism.*

COROLLARY A.16. *Let* $f : C_{*,*} \to \bar{C}_{*,*}$ *be a map of bicomplexes as above. Suppose that for each* $q \geq 0, H_p(C_{*,q}) = 0$ *and* $H_p(\bar{C}_{*,q}) = 0$ *for all* $p > 0$. *Let* $e : C_* \to C_{-1,*}$ *and* $\bar{e} : \bar{C}_* \to \bar{C}_{-1,*}$ *be the edge maps as in proposition A.13. Suppose further that for each* $p \geq 0$ *the map*

$$(\text{A.17}) \qquad\qquad f_* : H(C_{p,*}, \partial'') \to H(\bar{C}_{p,*}, \partial'')$$

is an isomorphism. Then there is a commutative diagram of isomorphisms

$$
\begin{array}{ccc}
H(C_*) & \xrightarrow{\;\; f_* \;\;}_{\cong} & H(\bar{C}_*) \\
e_* \downarrow \cong & & \cong \downarrow \bar{e}_* \\
H(C_{-1,*}) & \xrightarrow{\;\; f_* \;\;}_{\cong} & H(\bar{C}_{-1,*})
\end{array}
$$

Proof: This follows from corollary A.15 with $t = 1$ together with proposition A.13.

Remark 1: If $C_{p,*} = 0$ and $\bar{C}_{p,*} = 0$ for $p > 1$ then the above corollary is just the "five lemma" for the long exact sequences associated to the diagram with exact rows

$$
\begin{array}{ccccccccc}
0 & \longleftarrow & C_{-1,*} & \longleftarrow & C_{0,*} & \longleftarrow & C_{1,*} & \longleftarrow & 0 \\
& & \downarrow f & & \downarrow f & & \downarrow f & & \\
0 & \longleftarrow & \bar{C}_{-1,*} & \longleftarrow & \bar{C}_{0,*} & \longleftarrow & \bar{C}_{1,*} & \longleftarrow & 0
\end{array}
$$

Remark 2: We will sometimes use spectral sequences and bicomplexes with different gradings. The above theorems are then still valid when suitable adapted.

Now let us apply the above for the calculation of the chain complex $C_*(X)$ considered in chapter 2. In general for X any set let $\bar{C}_*(X)$ denote the chain-complex where a k-simplex, $k = 0, 1, 2, \ldots$, is any $(k+1)$-tuple $(a_0, \ldots, a_k), a_i \in X$ and the boundary map is given by

$$(A.18) \qquad \partial(a_0, \ldots, a_k) = \sum_{i=0}^{k}(-1)^i(a_0, \ldots, \hat{a}_i, \ldots, a_k).$$

Recall that by lemma 3.6 we have

PROPOSITION A.19. *For X any set the complex $\bar{C}_*(X)$ is acyclic.*

Now let X be any Riemannian manifold and suppose that there is a $\delta \in \mathbb{R}_+ \cup \{\infty\}$ such that every point $p \in X$ has a *normal neighborhood* U_p which is *geodesically convex* (See e.g. [**Helgason, 1962**, chapter I, §9]). That is, for each $p \in X$ the exponential map $\exp_p$ is a diffeomorphism from the open ball of radius δ in T_pX to U_p, and whenever $q, q' \in U_p$ there is a unique geodesic from q to q' contained in U_p. In particular δ exists if X is compact (e.g. $\delta = \pi/2$ for $X = S^n$). Also if X is a complete simply connected manifold of non-positive sectional curvature then we can take $\delta = \infty$ (e.g. for $X = E^n$ or $\mathcal{H}^n$). Now let $C_*^{\delta}(X)$ denote the subcomplex of $\bar{C}_*(X)$ of simplices $(a_0, \ldots, a_k)$ such that all $a_0, \ldots, a_k$ are contained in a normal neighborhood of radius δ. In particular $C_*^{\infty}(X) = \bar{C}_*(X)$. There is a natural map of $C_*^{\delta}(X)$ to the singular complex $C_*^{\mathrm{sing}}(X)$, where a k-simplex is a continuous map $f : \Delta^k \to X$ with the usual boundary map. Thus let $i(a_0, \ldots, a_k) = f$ be the map constructed inductively as the geodesic cone on the face $i(a_1, \ldots, a_k)$ with cone point a_0. Then

$$(A.20) \qquad i : C_*^{\delta}(X) \to C_*^{\mathrm{sing}}(X)$$

is clearly a chain map (we have used a different but equivalent construction in the classical geometries above, cf. proof of theorem 2.10).

PROPOSITION A.21. *The map i induces an isomorphism*

$$i_* : H(C_*^{\delta}(X)) \cong H(C_*^{\mathrm{sing}}(X)) = H_*(X, \mathbb{Z})$$

Proof: Let $\mathcal{U} = \{U_j, j \in J\}$ be the covering of X by all normal neighborhoods of radius δ. Consider the two bicomplexes

$$A_{p,q} = \bigoplus_{(j_0,\dots,j_p)} C_q^\delta(U_{j_0} \cap \dots \cap U_{j_p})$$

$$\bar{A}_{p,q} = \bigoplus_{(j_0,\dots,j_p)} C_q^{\text{sing}}(U_{j_0} \cap \dots \cap U_{j_p})$$

where we take the direct sum over all $(p+1)$-tuples of elements of J. In both bicomplexes $\partial'' = (-1)^p \partial$ where ∂ is the internal differential in C_*^δ respectively C_*^{sing} and

$$\partial' = \sum_{k=0}^{p} (-1)^k \epsilon_{k*}$$

where $\epsilon_{k*}, k = 0, \dots, p$, are induced by the inclusions

$$U_{j_0} \cap \dots \cap U_{j_p} \xrightarrow{\iota} U_{j_0} \cap \dots \cap \hat{U}_{j_k} \cap \dots \cap U_{j_p}.$$

Now each $C_*^\delta(U_{j_0} \cap \dots \cap U_{j_p}) = \bar{C}_*(U_{j_0} \cap \dots \cap U_{j_p})$ is either 0 (if $U_{j_0} \cap \dots \cap U_{j_p} = \emptyset$) or acyclic (if $U_{j_0} \cap \dots \cap U_{j_p} \neq \emptyset$) by proposition A.19. The same is true for $C_*^{\text{sing}}(U_{j_0} \cap \dots \cap U_{j_p})$ since $U_{j_0} \cap \dots \cap U_{j_p}$ is geodesically convex and hence contractible if it is non-empty. It follows that $i : A_{p,q} \to \bar{A}_{p,q}$ induces an isomorphism

(A.22) $i_* : H(A_{p,*}, \partial'') \xrightarrow[\cong]{} H(\bar{A}_{p,*}, \partial''), \quad p = 0, 1, 2, \dots$

Now put

$$A_{-1,*} = C_*^\delta(X), \quad \bar{A}_{-1,*} = C_*^{\text{sing}}(\mathcal{U})$$

where the latter denotes the subcomplex of $C_*^{\text{sing}}(X)$ generated by simplices contained in a neighborhood from $\mathcal{U}$. Then there are natural maps

$$\partial' : A_{0,*} \to A_{-1,*} \quad \text{and} \quad \partial' : \tilde{A}_{0,*} \to \tilde{A}_{-1,*}$$

extending the bicomplexes to $p \geq -1$, and also $i : A_{*,*} \to \tilde{A}_{*,*}$ extends in this range. By corollary A.16 and (A.22) above it suffices to show that

(A.23) $H_p(A_{*,q}) = 0$ and $H_p(\bar{A}_{*,q}) = 0 \quad \forall p \geq -1, q \geq 0.$

In fact, then we conclude that

$$i_* : H(C_*^\delta(X)) \xrightarrow[\cong]{} H(C_*^{\text{sing}}(\mathcal{U}))$$

is an isomorphism, and the latter is isomorphic to $H(C_*^{\mathrm{sing}}(X))$ by the excision theorem for singular homology (see e.g. [**Spanier, 1966**, chapter 4, §4]). To prove (A.23) let us do it for $A_{*,*}$; the proof for $\bar{A}_{*,*}$ is similar. We define a chain homotopy

$$s_p : A_{p,q} \to A_{p+1,q} \quad p = -1, 0, 1, \ldots$$

as follows: For $\sigma = (a_0, \ldots, a_q) \in C_q^\delta(X)$ choose $U_{j_\sigma} \in \mathcal{U}$ such that $a_i \in U_{j_\sigma}, i = 0, 1, \ldots, q$, and define s_p restricted to $C_*^\delta(U_{j_0} \cap \cdots \cap U_{j_p})$ by

$$s_p(\sigma_{j_0 \ldots j_p}) = \sigma_{j_\sigma j_0 \ldots j_p} \in C_*^\delta(U_{j_\sigma} \cap \cdots \cap U_{j_p}), \quad p = -1, 0, 1, \ldots$$

Then one checks that

$$\partial' s_p + s_{p-1}\partial' = \mathrm{id}, \quad p = -1, 0, 1, \ldots$$

which proves that $H_p(A_{*,q}) = 0, \forall p \geq -1, \forall q \geq 0$. This ends the proof. $\square$

Another useful application of bicomplexes is the *hyperhomology spectral sequence* for a chain complex (N_*, ∂) of (left) $R[G]$-modules for G a group and R a commutative ring with unit. Assume for convenience that $N_i = 0$ for $i < 0$. As in (13.3) we define the *hyperhomology* by

(A.24) $$\boldsymbol{H}_n(G, N_*) = H_n(G \backslash (B_*(G) \otimes_R N_*))$$

i.e. the homology of the total complex of the bicomplex

(A.25) $$A_{p,q} = G \backslash (B_p(G) \otimes_R N_q),$$

where $(B_*(G), \partial_G)$ is the bar complex defined in chapter 4. Then the "second" spectral sequence in (A.11-12) gives (with transposed indices) a spectral sequence

(A.26) $$E_{p,q}^r = {}^{"}E_{q,p}^r \Rightarrow \boldsymbol{H}_{p+q}(G, N_*)$$

where

(A.27) $$E_{p,q}^1 = H_p(G, N_q)$$

and

(A.28) $$E_{p,q}^2 = H_q(H_p(G, N_*), \partial_*)$$

COROLLARY A.29. *Suppose the complex N_* is acyclic with $H_0(N_*) = M$. Then there is a natural isomorphism*

$$\boldsymbol{H}_n(G, N_*) \cong H_n(G, M)$$

In particular the hyperhomology spectral sequence (A.26) converges to $H_*(G, M)$.

Proof: The bicomplex $A_{*,*}$ in (A.25) has for all p,

$$H_q(A_{p,*}) = \begin{cases} G\backslash(B_p(G) \otimes M) & q = 0 \\ 0 & q > 0. \end{cases}$$

Then the statement of the corollary follows from proposition A.13. □

Bibliography

[Akin-Buchsbaum-Weyman, 1982]   K. Akin, D. A. Buchbaum, and J. Weyman, *Schur functors and Schur complexes,* Adv. Math. 44 (1982), 207-278.

[Beilinson et. al, 1991]   A. A. Beilinson, A. B. Goncharov, V. V. Schechtman and A. N. Varchenko, *Aomoto dilogarithms, mixed Hodge structures and motivic cohomology of pairs of triangles in the plane,* in Grothendieck Festschrift II, pp. 78-131, Progress in Math. 87, Birkhäuser, Boston, 1991.

[Bloch, 1977]   S. J. Bloch, *Applications of the dilogarithm function in algebraic K-theory and algebraic geometry,* Proc. Int. Symp. Alg. Geom., pp. 1-14, Kyoto, 1977.

[Bloch, 1978]   S. J. Bloch, *Higher Regulators, Algebraic K-theory, and Zeta Functions of Elliptic Curves,* Irvine Lecture Notes, 1978. (Unpublished.)

[Boltianskii, 1978]   V. G. Boltianskii, *Hilbert's Third Problem,* Wiley, New York, 1978.

[Borel, 1977]   A. Borel, *Cohomologie de SL_n et valeurs de fonctions zeta aux points entiers,* Ann. Sci. Norm. Sup. Pisa, Cl. Sci. 4 (1977), 613-616, correction, ibid. 7 (1980) 373. See also Collected Papers, 3 (1983), 495-519.

[Bricard, 1896]   R. Bricard, *Sur une question de géométrie relative aux polyèdres,* Nouv. Ann. Math., 15 (1896), 331-334.

[Bökstedt-Brun-Dupont, 1998]   M. Bökstedt, M. Brun, and J. L. Dupont, *Homology of $O(n)$ and $O^1(1,n)$ made discrete: an application of edgewise subdivision,* J. Pure Appl. Algebra, 123 (1998), 131-152.

[Cartan-Eilenberg, 1956]   H. Cartan and S. Eilenberg, *Homological Algebra,* Princeton Univ. Press, 1956.

[Cartier, 1985]   P. Cartier, *Décomposition des polyèdres: le point sur le troisième problème de Hilbert*, Sém. Bourbaki, 37ème année, 1984-85, no. 646.

[Cathelineau, 1992]   J.-L. Cathelineau, *Quelques aspects du troisième problème de Hilbert*, Gaz. Math. 52 (1992), 45-71.

[Cathelineau, 1993]   J.-L. Cathelineau, *Homologie du groupe linéaire et polylogarithmes (d'après A. B. Goncharov et d'autres)*, Séminaire Bourbaki, 1992/93, Astérisque 216 (1993), exp. 772, 5, 311-341.

[Cathelineau, 1995]   J.-L. Cathelineau, *Birapport et groupoïdes*, Enseign. Math. 41 (1995), 257-280.

[Cathelineau, 1996]   J.-L. Cathelineau, *Remarques sur les différentielles des polylogarithmes uniformes*, Ann. Inst. Fourier (Grenoble) 46 (1996), 1327-1347.

[Cathelineau, 1998]   J.-L. Cathelineau, *Homology of tangent groups considered as discrete groups and scissors congruence*, J. Pure Appl. Algebra, 132 (1998), 9-25.

[Cheeger, 1974]   J. Cheeger, *Invariants of flat bundles*, Proc. Int. Cong. of Math., Vancouver, (1974), 3-6.

[Cheeger-Simons, 1985]   J. Cheeger and J. Simons, *Differential characters and geometric invariants*, in Geometry and Topology, eds. J. Alexander and J. Harer, pp. 50-80, Lecture Notes in Math. 1167, Springer-Verlag, Berlin, 1985.

[Chern-Simons, 1974]   S.-S. Chern and J. Simons, *Characteristic forms and geometric invariants*, Ann. of Math. 99 (1974), 48-69.

[Coxeter, 1935]   H. S. M. Coxeter, *The functions of Schläfli and Lobatschefsky*, Quart. J. Math., 6 (1935), 13-29.

[Coxeter, 1973]   H. S. M. Coxeter, *Regular Polytopes*, 3rd ed., Dover, 1973.

[Dehn, 1901]   M. Dehn, *Über den Rauminhalt*, Math. Ann., 105 (1901), 465-478.

[Dehn, 1905]   M. Dehn, *Über den Inhalt spharischer Dreiecke*, Math. Ann., 60 (1905), 166-174.

[Dehn, 1906]   M. Dehn, *Die Eulersche Formel im Zusammenhang mit dem Inhalt in der Nicht-Euklidischen Geometrie*, Math. Ann., 61 (1906), 561-586.

[Dupont, 1976]   J. L. Dupont, *Simplicial de Rham cohomology and characteristic classes for flat bundles*, Topology, 15(1976), 233-245.

[Dupont, 1978]   J. L. Dupont, *Curvature and Characteristic Classes*, Lecture Notes in Math. 640, Springer-Verlag, Berlin, 1978.

[Dupont, 1982]   J. L. Dupont, *Algebra of polytopes and homology of flag complexes*, Osaka J. Math. 19 (1982), 599-641.

[Dupont, 1987]   J. L. Dupont, *The dilogarithm as a characteristic class for flat bundles*, J. Pure Appl. Algebra, 44 (1987), 137-164.

[Dupont-Kamber, 1990]   J. L. Dupont and F. W. Kamber, *On a generalization of Cheeger-Chern-Simons classes*, Ill. J. Math. 34 (1990), 221-255.

[Dupont-Kamber, 1993]   J. L. Dupont and F. W. Kamber, *Cheeger-Chern-Simons classes of transversally symmetric foliations: Dependence relations and eta-invariants*, Math. Ann. 295 (1993), 449-468.

[Dupont-Parry-Sah, 1988] J. L. Dupont, W. Parry and C.-H. Sah, *Homology of classical Lie groups made discrete, II*, J. Algebra 113 (1988), 215-260.

[Dupont-Sah, 1982]   J. L. Dupont and C.-H. Sah, *Scissors congruences, II*, J. Pure and Appl. Algebra, 25 (1982), 159-195. Corrigendum, 30 (1983), 217.

[Dupont-Sah, 1990]   J. L. Dupont and C.-H. Sah, *Homology of Euclidean groups of motions made discrete and Euclidean scissors congruences*, Acta Math. 164 (1990), 1-27.

[Dupont-Sah, 1994]   J. L. Dupont and C.-H. Sah, *Dilogarithm identities in conformal field theory and group homology*, Commun. Math. Phys. 161 (1994), 265-282.

[Dupont-Sah, 1999]   J. L. Dupont and C.-H. Sah, *Three questions about simplices in spherical and hyperbolic 3-space*, in The Gelfand Mathematical Seminars, 1997-1999, pp. 50-76, eds. I. M. Gelfand and V. Retakh, Birkhäuser, Boston, 1999.

[Gauss, 1844]   C. F. Gauss, *Werke, Bd. 8*, p. 241, p. 244, Teubner, Leipzig, 1900.

[Goncharov, 1995]   A. B. Goncharov, *Geometry of configurations, polylogarithms, and motivic cohomology*, Adv. Math. 114 (1995), 197-318.

[Goncharov, 1999]   A. B. Goncharov, *Volumes of hyperbolic manifolds and mixed Tate motives*, J. Amer. Math. Soc. 12 (1999), 569-618.

[Haagerup-Munkholm, 1981] U. Haagerup and H. Munkholm, *Simplices of maximal volume in hyperbolic n-space*, Acta Math. 147 (1981), 1-11.

[Hadwiger, 1957] H. Hadwiger, *Vorlesungen über Inhalt, Oberfläche und Isoperimetrie*, Springer-Verlag, Berlin 1957.

[Hadwiger, 1967] H. Hadwiger, *Neuere Ergebnisse innerhalb der Zerlegungstheorie euclidischer Polyeder*, Jber. Deutsch. Math.-Verein. 70 (1967/68), 167-176.

[Hadwiger, 1968] H. Hadwiger, *Translative Zerlegungsgleichheit der Polyeder des gewohnlicher Raumes*, J. Reine Angew. Math., 233 (1968), 200-212.

[Hain, 1994] R. M. Hain, *Classical polylogarithms*, in *Motives* pp. 3-42, Proc. Symp. Pure Math., 55, Amer. Math. Soc., Providence, R. I., 1994.

[Hain-Yang, 1996] R. M. Hain and J. Yang, *Real Grassmann polylogarithms and Chern classes*, Math. Ann. 304 (1996), 157-201.

[Helgason, 1962] S. Helgason, *Differential Geometry and Symmetric Spaces*, Academic Press, New York, 1962.

[Hilbert, 1900] D. Hilbert, *Gesammelte Abhandlungen, Bd. 3*, pp. 301-302, Chelsea, 1965.

[Hirzebruch, 1966] F. Hirzebruch, *Topological Methods in Algebraic Geometry*, Grundlehren Math. Wiss. 131, Springer-Verlag, Berlin, 1966.

[Hochschild-Kostant-Rosenberg, 1962] G. Hochschild, B. Kostant and A. Rosenberg, *Differential forms on regular affine algebras*, Trans. Amer. Math. Soc. 102 (1962), 383-408.

[Hopf, 1983] H. Hopf, *Differential Geometry in the Large*, Lecture Notes in Math. 1000, Springer-Verlag, Berlin, 1983.

[Iversen, 1992] B. Iversen, *Hyperbolic Geometry*, London Math. Soc. Student Texts 25, Cambridge Univ. Press, Cambridge, 1992.

[Jackson, 1912] W. H. Jackson, *Wallace's theorem concerning plane polygons of the same area*, Amer. J. Math. 34 (1912), 383-390.

[Jessen, 1941] B. Jessen, *A remark on the volume of polyhedra*, Mat. Tidsskr. B (1941), 59-65. (In Danish.)

[Jessen, 1968] B. Jessen, *The algebra of polyhedra and the Dehn-Sydler theorem*, Math. Scand., 22 (1968), 241-256.

[Jessen, 1972] B. Jessen, *Zur Algebra der Polytope*, Göttingen Nachr. Math. Phys. (1972), 47-53.

[Jessen, 1973] B. Jessen, *Polytope algebra*, in *Danish Mathematical Society* 1923-1973, ed. F. D. Pedersen, pp. 83-92, Danish Math. Soc., Copenhagen, 1973. (In Danish.)

[Jessen, 1978]   B. Jessen, *Einige Bermerkungen zur Algebra der Polyeder in nicht-Euclideschen Räumen*, Comment. Math. Helv., 53 (1978), 424-529

[Jessen-Karpf-Thorup, 1968]   B. Jessen, J. Karpf and A. Thorup, *Some functional equations in groups and rings*, Math. Scand. 22 (1968), 257-265.

[Jessen-Thorup, 1978]   B. Jessen and A. Thorup, *The algebra of polytopes in affine spaces*, Math. Scand. 43 (1978), 211-240.

[Kamber-Tondeur, 1968]   F. W. Kamber and Ph. Tondeur, *Flat Manifolds*, Lecture Notes in Math. 67, Springer-Verlag, Berlin, 1968.

[Karoubi, 1986]   M. Karoubi, *K-théorie multiplicative et homologie cyclique*, C. R. Acad. Sci. Paris Sér. A-B, 303 (1986), 507-510.

[Kassel, 1982]   C. Kassel, *La K-théorie stable*, Bull. Soc. Math. France, 110 (1982), 381-416.

[Koszul, 1968]   J. L. Koszul, *Lecture on hyperbolic Coxeter groups*, Univ. of Notre Dame, 1967/68.

[Lefschetz, 1953]   S. Lefschetz, *Algebraic Geometry*, Princeton Math. Series 18, Princeton Univ. Press, Princeton, N. J., 1953.

[Lewin, 1981]   L. Lewin, *Polylogarithms and Associated Functions*, Elsevier, 1981.

[Lewin, 1991]   L. Lewin, ed., *Structural Properties of Polylogarithms*, Math. Surveys and Monographs, 37, Amer. Math. Soc., Providence, 1991.

[Lusztig, 1974]   G. Lusztig, *The Discrete Series of GL_n over a Finite Field*, Ann. of Math. Studies 81, Princeton University Press, Princeton, N.J., 1974.

[MacLane, 1963]   S. MacLane, *Homology*, Grundlehren Math. Wiss. 114, Springer-Verlag, Berlin 1963.

[May, 1967]   J. P. May, *Simplicial Objects in Algebraic Topology*, D. Van Nostrand Co., Toronto, 1967.

[Milnor, 1980]   J. W. Milnor, *How to compute volume in hyperbolic space*, Collected Papers, I, Publish or Perish, 1994.

[Milnor, 1982]   J. W. Milnor, *Hyperbolic Geometry: The first 150 years*, Collected Papers, I, Publish or Perish, 1994.

[Milnor, 1983]   J. W. Milnor, *On the homology of Lie groups made discrete*, Comment. Math. Helv. 58 (1983), 72-85.

[Morelli, 1993]   R. Morelli, *Translation scissors congruence*, Adv. Math., 100 (1993), 1-27.

[Mumford, 1976]   D. Mumford, *Algebraic Geometry I: Complex Projective Varieties*, Grundlehren Math. Wiss. 221, Springer-Verlag, Berlin, 1976.

[Neumann, 1998]   W. D. Neumann, *Hilbert's 3rd problem and invariants of 3-manifolds* in *The Epstein birthday schrift*, pp. 383-411 (eletronic), Geom. Topol. Monogr., 1, Geom. Topol., Coventry, 1998.

[Neumann-Yang, 1995]   W. D. Neumann and J. Yang, *Problems for K-theory and Chern-Simons invariants of hyperbolic 3-manifolds*, L'Ens. Math. 41 (1995), 281-296.

[Neumann-Yang, 1995]   W. D. Neumann and J. Yang, *Invariants from triangulation for hyperbolic 3-manifolds*, Elec. Res. Announce. of Amer. Math. Soc. 1 (1995), 72-79.

[Neumann-Yang, 1999]   W. D. Neumann and J. Yang, *Bloch invariants of hyperbolic 3-manifolds*, Duke Math. J. 96(1999), 29-59.

[Neumann-Zagier, 1985]   W. D. Neumann and D. Zagier, *Volumes of hyperbolic three-manifolds*, Topology, 24 (1985), 307-332.

[Raghunatan, 1972]   M. S. Raghunatan, *Discrete Subgroups of Lie Groups*, Ergebnisse Math. 68, Springer-Verlag, Berlin, 1972.

[Reznikov, 1995]   A. Reznikov, *All regulators of flat bundles are torsion*, Ann. of Math. 141 (1995), 373-386.

[Roger, 1975]   C. Roger, *Sur l'homologie des groupes de Lie*, C. R. Acad. Sci. Paris Sér. A-B, 280 (1975), A325-327.

[Sah, 1979]   C. H. Sah, *Hilbert's Third Problem: Scissors Congruence*, Res. Notes in Math. 33, Pitman, 1979.

[Sah, 1981]   C. H. Sah, *Scissors congruence, I, Gauss-Bonnet map*, Math. Scand. 49 (1981), 181-210.

[Sah, 1986]   C. H. Sah, *Homology of Lie groups made discrete, I*, Comment. Math. Helv. 61 (1986), 308-347.

[Sah, 1989]   C. H. Sah, *Homology of Lie groups made discrete, III*, J. Pure and Appl. Algebra, 56 (1989), 269-312.

[Sah-Wagoner, 1977]   C. H. Sah and J. B. Wagoner, *Second homology of Lie groups made discrete*, Comm. in Algebra, 5 (1977), 611-642.

[Schläfli, 1860]   L. Schläfli, *On the multiple integral...*, Quart J. Math. 3 (1860), 54-68, 97-108, see also Gesam. Math. Abhånd., 1, Birkhauser, 1950.

[Spanier, 1966]   E. H. Spanier, *Algebraic Topology*, MacGraw-Hill, New York, 1966.

[Suslin, 1983]   A. A. Suslin, *On K-theory of algebraically closed fields*, Invent. Math. 73 (1983), 241-245.

[Suslin, 1984]   A. A. Suslin, *Homology of GL_n, characteristic classes and Milnor K-theory*, in *Algebraic K-theory, Number Theory, Geometry and Analysis*, ed. A. Bak, pp. 357-375, Lecture Notes in Math. 1046, Springer-Verlag, Berlin, 1984.

[Suslin, 1986]   A. A. Suslin, *Algebraic K-theory of fields*, Proc. Int. Cong. Math., I, Berkeley, (1986), 222-244.

[Suslin, 1991]   A. A. Suslin, *K_3 of a field and the Bloch group*, Proc. Steklov Inst. of Math., (1991), 217-239.

[Sydler, 1965]   J. P. Sydler, *Conditions nécessaires et suffisantes pour l'equivalence des polyèdres de l'espace euclidien à trois dimensions*, Comment. Math. Helv., 40 (1965), 43-80.

[Thurston, 1979]   W. P. Thurston, *The Geometry and Topology of Three-Manifolds,* Geometry Center, Univ. of Minn., 1979.

[Wallace, 1807]   W. Wallace, *Question 269,* in Thomas Leyborne, Math. Repository, III, London, 1814.

[Whitney, 1938]   H. Whitney, *Tensor products of abelian groups,* Duke Math. J., 4 (1938), 495-528.

[Yoshida, 1985]   T. Yoshida, *The η-invariant of hyperbolic 3-manifolds,* Invent. Math. 81 (1985), 437-514.

[Zylev, 1965]   V. B. Zylev, *Equicomposability of equicomplementable polyhedra,* Sov. Math. Doklady, 161 (1965), 453-455.

[Zylev, 1968]   V. B. Zylev, *G-composedness and G-complementability,* Sov. Math. Doklady, 179 (1968), 403-404.

Index

$2(n + 1)$-gon, 136
G-scissors congruent, 5
η-invariant, 102
k-wedge, 138

affine subspace, 17
Alexander-Whitney map, 33
algebraic K-theory, 7
alternating chain complex, 125

bicomplex, 152
Bloch conjecture, 103
Bloch-Wigner function, 96

Cheeger-Chern-Simons classes, 91, 92
circumcenter, 81, 82
cone, 56
continuous cochain, 100
convergence of spectral sequence, 152
corner, 138

decomposable configurations, 126, 135
decomposition, 4
degeneracy operators, 17
Dehn invariant, 3
differential graded algebra, 47
dilogarithmic function, 91, 96

dimension filtration, 12
dual simplex, 109, 110

edgewise subdivision, 84
Eilenberg-Zilber map, 28
even sign changes, 111

face operators, 17
filtration, 151
flag, 18
flat connection, 92
flat simplex, 12

general position, 80
geodesic cone, 94
geodesic simplex, 94
geodesically convex, 4, 155
geometric n-simplex, 4
Gram matrix, 109, 110
group homology, 6, 27

Hadwiger invariant, 34
Hochschild homology, 46
homogeneous bar complex, 27
homology, 5
hyperbolic orthoscheme, 113
hyperbolic simplex, 83, 110

hyperbolic subspace, 17
hyperhomology, 126, 157
hyperhomology spectral sequence, 157

ideal vertices, 73
incident, 34
independent configurations, 127
inhomogeneous bar complex, 28
irreducible components, 122

Kähler differentials, 45

length of flag, 18
lune, 60
Lusztig exact sequences, 25

non-commutative differential forms, 45,
    47
non-degenerate subspace, 107
normal neighborhood, 155

odd sign changes, 111
orientation module, 21
orthoscheme, 113

polytope module, 14
Pontrjagin product, 29
prism, 32
projective configurations, 7, 125
proper simplex, 12
proper subspace, 18, 30

rank filtration, 12
rational lune, 60
rational orthoschemes, 115
rational simplex, 112
regulator, 101
representation variety, 121
rigidity, 120
Rogers' L-function, 99

Schur functor, 132
scissors congruence group, 5
scissors congruent (s.c.), 1
secondary characteristic class, 7
shuffle product, 47
similarities, 66
simplex, 12

simplicial abelian group, 46
simplicial set, 17
spectral sequence, 151
spherical orthoscheme, 113
spherical simplex, 109
spherical subspace, 17
Steinberg module, 15
strict flag, 18, 20
suspension, 54, 110
suspension homomorphism, 24

Tits complex, 18
total complex, 126, 152
translational scissors congruence group,
    27
translations, 27
twisted Cheeger-Chern-Simons invari-
    ant, 101

vertices, 4